NELSON
VICmaths

VCE UNITS ① + ②

T0358031

specialist
mathematics 11

mastery workbook

Greg Neal
Sue Garner
George Dimitriadis
Stephen Swift

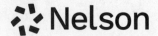

Nelson VICmaths Specialist Mathematics 11 Mastery Workbook
1st Edition
Greg Neal
Sue Garner
George Dimitriadis
Stephen Swift
ISBN 9780170464109

Publisher: Dirk Strasser
Additional content created by: ansrsource
Project editor: Alan Stewart
Series cover design: Leigh Ashforth (Watershed Art & Design)
Series text design: Rina Gargano (Alba Design)
Series designer: Nikita Bansal
Production controller: Karen Young
Typeset by: MPS Limited

Any URLs contained in this publication were checked for currency during the production process. Note, however, that the publisher cannot vouch for the ongoing currency of URLs.

Acknowledgements

TI-Nspire: Images used with permission by Texas Instruments, Inc
Casio ClassPad: Shriro Australia Pty. Ltd.

© 2022 Cengage Learning Australia Pty Limited

For product information and technology assistance,
in Australia call **1300 790 853**;
in New Zealand call **0800 449 725**

For permission to use material from this text or product, please email
aust.permissions@cengage.com

ISBN 978 0 17 046410 9

Cengage Learning Australia
Level 5, 80 Dorcas Street
Southbank VIC 3006 Australia

Cengage Learning New Zealand
Unit 4B Rosedale Office Park
331 Rosedale Road, Albany, North Shore 0632, NZ

For learning solutions, visit **cengage.com.au**

Printed in China by 1010 Printing International Limited.
1 2 3 4 5 6 7 26 25 24 23

Contents

Matrices   **110**

5

Counting techniques   **128**

6

Trigonometric identities

7

Graphing functions and relations

To the student

Nelson VICmaths is your best friend when it comes to studying Specialist Mathematics in Year 11. It has been written to help you maximise your learning and success this year. Every explanation, every exam hack and every worked example has been written with the exams in mind.

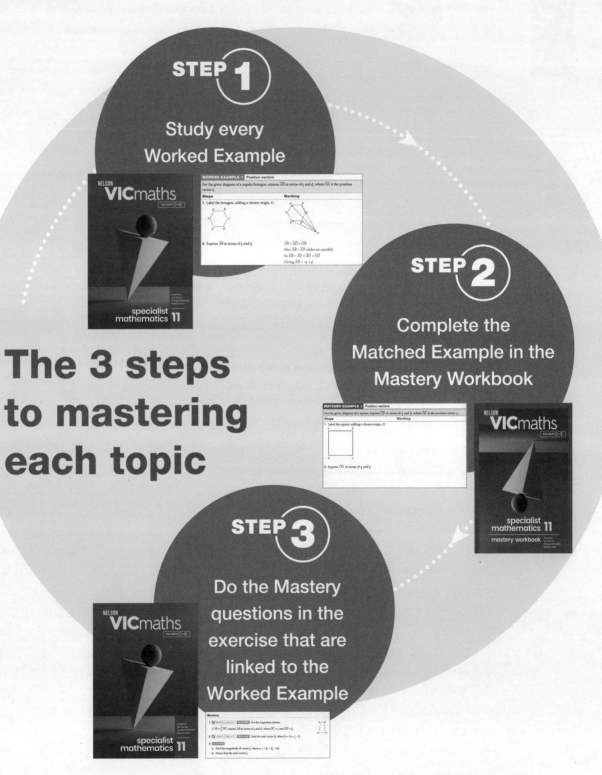

The 3 steps to mastering each topic

STEP 1
Study every Worked Example

STEP 2
Complete the Matched Example in the Mastery Workbook

STEP 3
Do the Mastery questions in the exercise that are linked to the Worked Example

SEQUENCES AND SERIES

CHAPTER

1

MATCHED EXAMPLE 1	Recurrence relations

Find the recurrence relation for the sequence 2, 4, 8, 16, …

SB

p. 4

Steps	Working
1 State t_1.	
2 Find a pattern for the terms.	
3 Write the recurrence relation.	

SB

Using CAS 1:
Generating
sequences
p. 5

Using CAS 2:
Recurrence
relations
p. 6

MATCHED EXAMPLE 2 | Finding terms in a sequence

Find the first 5 terms for the sequence defined by the rule $t_n = n^2 - n$, where n is a natural number.

Steps	Working
1 State t_1.	
2 Substitute values of n to write the terms.	
3 Write the first 5 terms for the sequence.	

MATCHED EXAMPLE 3 | Arithmetic sequence

Determine if each sequence is arithmetic, and if so, write the next 4 terms.

a $2, 5, 8, 11, \ldots$ **b** $2, 2+\sqrt{3}, 2+2\sqrt{3}, 2+3\sqrt{3}, \ldots$ **c** $3, 2, -2, -10, \ldots$

Steps	Working
a **1** Evaluate $t_2 - t_1$.	
2 Evaluate $t_3 - t_2$.	
3 Evaluate $t_4 - t_3$.	
4 The difference is always 3.	
Each term is 3 more than the previous one.	
b **1** Evaluate $t_2 - t_1$ and $t_3 - t_2$ and $t_4 - t_3$.	
2 Each term is $\sqrt{3}$ more than the previous one.	
c **1** Evaluate $t_2 - t_1$ and $t_3 - t_2$ and $t_4 - t_3$.	
2 The decreases from one term to the next are different.	

MATCHED EXAMPLE 4 Terms in an arithmetic sequence

Find the 17th term of the arithmetic sequence 2, 7, 12, …

Steps	Working
1 Write the first term, a.	
2 Find the common difference, d.	
3 Write the number of terms, n.	
4 Substitute the values into the formula $t_n = a + (n-1)d$.	

MATCHED EXAMPLE 5 | Finding a term of an arithmetic sequence

Which term of the arithmetic sequence 6, 10, 14, … is equal to 138?

Steps	**Working**
1 Write the first term, a, and the common difference, d.	
2 Write the formula for the nth term and make it equal to 138.	
3 Substitute the values into the formula and solve for n.	
4 Write the answer.	

SB

p. 12

MATCHED EXAMPLE 6 | Finding an arithmetic sequence

Find the 1st 5 terms of the arithmetic sequence with a 4th term of 12 and a 15th term of 45.

Steps	Working
1 Write t_4 in the form $a + (n-1)d$.	
2 Write t_{15} in the form $a + (n-1)d$.	
3 Solve simultaneously to find a and d.	

TI-Nspire   ClassPad

4 Use $a = 3$ and $d = 3$ to write the AP.

MATCHED EXAMPLE 7	Arithmetic series

Evaluate the arithmetic series $5 + 8 + 11 + 14 + \ldots$ for 12 terms.

Steps	Working
1 Write down the information provided.	
2 Evaluate the sum of the 12 terms using the formula for S_n.	

MATCHED EXAMPLE 8	Terms in an arithmetic series

If $7 + 11 + 15 + \ldots + t_n = 250$, how many terms are there in the series?

Steps	Working
1 Write down the information provided.	
2 Substitute values into S_n to find n.	
3 Solve the equation for n.	
4 n must be a natural number.	

MATCHED EXAMPLE 9 | Finding an arithmetic series

The 3rd term of an arithmetic series is 21 and the sum of the first 8 terms is 240.

Find the sum of the first 17 terms.

Steps	Working
1 Write down the information provided.	
2 Substitute values into suitable formulas.	
3 Use CAS to solve simultaneous equations [1] and [2] to find a and d.	
TI-Nspire	**ClassPad**
4 Calculate the value of S_{17} using S_n.	

MATCHED EXAMPLE 10 | Common ratio of geometric sequence

Determine the common ratio for each geometric sequence.

a $1, 4, 16, 64,$ **b** $2, \dfrac{2}{3}, \dfrac{2}{9}, \dfrac{2}{27}$

Steps	Working
a 1 Evaluate $\dfrac{t_2}{t_1}$ and $\dfrac{t_3}{t_2}$ and $\dfrac{t_4}{t_3}$.	
2 Find the common ratio, r.	
b 1 Evaluate $\dfrac{t_2}{t_1}$ and $\dfrac{t_3}{t_2}$.	
2 Find the common ratio, r.	

MATCHED EXAMPLE 11	Terms of a geometric sequence

What is the 11th term in the geometric sequence 5, −20, 80, …?

Steps	Working
1 Write the first term, a.	
2 Find the common ratio, r.	
3 Write the number of terms, n.	
4 Substitute the values into the formula.	
5 Write the answer.	

MATCHED EXAMPLE 12 | Finding a term of a geometric sequence

Which term of the geometric sequence $5, -1, \dfrac{1}{5}, \ldots$ is equal to $-\dfrac{1}{244\,140\,625}$?

Steps	Working
1 Write the first term, a, and the common ratio, r.	
2 Substitute into the formula, make $t_n = -\dfrac{1}{244\,140\,625}$ and solve for n using CAS.	
3 Write the answer.	

9780170464109

MATCHED EXAMPLE 13 | Finding a geometric sequence

The 3rd and 5th terms of a geometric sequence are 63 and 567, respectively.

What is the 9th term of the sequence?

Steps	Working
1 Write t_3 and t_5 in the form $t_n = ar^{n-1}$ and set them equal to 63 and 567, respectively.	
2 Use CAS to solve these simultaneous equations to find a and r.	

TI-Nspire   **ClassPad**

3 Use $a = 7, r = 3$ or $a = 7, r = -3$ to find the 9th term of the sequence.

MATCHED EXAMPLE 14 | Convergence in a geometric sequence

Do the terms in the geometric sequence $\dfrac{1}{2}, \dfrac{3}{4}, \dfrac{9}{8}, \ldots$ converge?

Steps	Working
1 Find the common ratio using $r = \dfrac{t_2}{t_1} = \dfrac{t_3}{t_2}$.	
2 $r = \dfrac{3}{2}$ is greater than 1.	

MATCHED EXAMPLE 15	Geometric series

Evaluate the geometric series $\dfrac{2}{5} + \dfrac{4}{25} + \dfrac{8}{125} + \dfrac{16}{625} + \cdots$ for 9 terms.

Steps	Working
1 Write down the information provided.	
2 Evaluate the sum of the 9 terms using the formula for S_n.	

MATCHED EXAMPLE 16 | Number of terms in a geometric series

If $3 + 6 + 12 + \ldots + t_n = 12\,285$, how many terms are there in this geometric series?

Steps	Working
1 Write down the information provided.	
2 Substitute values into formula for S_n.	
3 Solve for n.	
4 Write the answer.	

MATCHED EXAMPLE 17 | Finding a term in a geometric series

The common ratio of a geometric series is 5 and the sum of the first 7 terms is 19 531.

Find the first term.

Steps	Working
1 Write down the information provided.	
2 Substitute values into $S_n = \dfrac{a(r^n - 1)}{r - 1}$.	
3 Solve for a.	
4 Write the answer.	

MATCHED EXAMPLE 18 | Infinite geometric series

Find the limiting sum of the infinite geometric series

$$\frac{5}{2} + \frac{5}{6} + \frac{5}{18} + \frac{5}{54} + \ldots$$

Steps	Working
1 Find a and r.	
2 Substitute values into the formula for S_∞ to find the limiting sum.	
3 Write the answer.	

MATCHED EXAMPLE 19	Recurring decimals

Express the recurring decimal $0.\dot{2}\dot{1}$ as a fraction.

p. 27

Steps	Working

1 Expand $0.\dot{2}\dot{1}$ as an infinite geometric series and find a and r.

2 Substitute values into $S_\infty = \dfrac{a}{1-r}$ to find the limiting sum.

3 Write the answer.

①

MATCHED EXAMPLE 20 | General term of a sequence

Find the rule for the nth term of the sequence 1, 8, 27, 64, 125, … in terms of n.

Steps	Working
1 Consider $n \in \{1,2,3,4,5\}$ and compare with the terms of the sequence.	
2 The terms are the cubes of n.	

MATCHED EXAMPLE 21 | Linear recurrence relation

Use the recurrence relation to find the first 5 terms of the sequence $t_1 = 1$ and $t_{n+1} = \dfrac{1}{3}t_n + 2.$

Steps	Working
1 The first term is given.	
2 Find t_{n+1} for $n = 2, 3, 4,$ and 5 using $t_{n+1} = \dfrac{1}{3}t_n + 2$.	
3 List the first 5 terms.	

MATCHED EXAMPLE 22 | The explicit rule of a linear recurrence relation

A sequence of numbers is defined by the linear recurrence relation $t_{n+1} = 5t_n - 2$, with $t_1 = 3$.

Find the explicit rule for t_n in terms of n only.

Steps	Working
1 Write t_1, t_2, t_3, t_4, t_5 using the rule to find a pattern, but do not calculate.	
2 Separate the parts of t_5.	
3 Factorise appropriately.	
4 Use the geometric series formula $S_n = \dfrac{a(r^n - 1)}{r - 1}$.	
5 Simplify.	
6 Generalise.	

SB

Using CAS 4:
Generating
linear recurrence
relations
p. 32

MATCHED EXAMPLE 23 | The explicit rule of a linear recurrence relation using CAS

A sequence of numbers is defined by the linear recurrence relation $t_{n+1} = 2 + 5t_n$ where $t_1 = 1$. Find the explicit rule for t_n in terms of n only.

SB
p. 39

Steps	Working
1 Calculate the values of t_1, t_2 and t_3 using the rule.	
2 The explicit rule is of the form $t_n = ka^{n-1} + c$. Define this function and solve for k, a and c. **TI-Nspire** **ClassPad**	
3 Write the answer.	

MATCHED EXAMPLE 24 | Compound interest calculation 1

An investment of $5000 earns 7.5% p.a. compound interest over 5 years.

a Write a recurrence relation for the amount at the end of each year.

b Use the relation to write an expression for the amount at the end of 5 years.

Steps	Working
a 1 Write the interest as a decimal.	
2 Write the growth factor.	
3 Write the recurrence relationship for A_n, the amount after n years.	
b 1 Write the amounts after the first 3 years.	
2 Generalise to A_n, the amount after n years.	
3 Write the expression for the amount after 5 years.	

MATCHED EXAMPLE 25 | Compound interest calculation 2

Find an expression for the value of an investment of $6500 under compound interest at 3% p.a. calculated monthly after 9 years 4 months.

Steps	Working
1 Write the monthly interest as a decimal.	
2 Write the growth factor.	
3 Write the recurrence relationship for A_n, the amount after n months.	
4 Write the amount after n months.	
5 Find n for the time 9 years 4 months.	
6 Write an expression for the value after 9 years and 4 months.	

MATCHED EXAMPLE 26	Return on a regular investment

Mark starts his earning journey at age 22 and puts \$3000 per year into a superannuation account earning 7% p.a. interest at the end of each year.

a Write a recurrence relation for his investment each year.

b Find an expression for the value of his investment after n years.

c Write an expression for the value of his investment when he turns 35.

Steps	Working
a Use A_n for the value at the beginning of the nth year.	
b 1 Write expressions for the first few values.	
2 Simplify the expression for A_4.	
3 Use the geometric series formula $$S_n = \frac{a(r^n - 1)}{r - 1}$$	
4 Generalise the expression.	
5 Write the answer.	
c When he turns 35 he will have been working for 13 years.	

MATCHED EXAMPLE 27 | Reducing balance loan

Jill borrowed $750 000 at 10.5% p.a. interest to buy a Bungalow, making payments of $85 000 a year.

a Write a recurrence relation for the amount left to pay after each year.

b Find an expression for the amount left to pay after n years.

c Write an expression for the amount left to pay after 7 years.

Steps	Working
a Use A_n for the amount left to pay at the end of the nth year.	
b 1 Write expressions for the first few values.	
2 Simplify the expression for A_4.	
3 Use the geometric series formula $$S_n = \frac{a(r^n - 1)}{r - 1}.$$	
4 Generalise the expression.	
5 Write the answer.	
c Find the amount after 7 years.	

MATCHED EXAMPLE 28 | Reducing balance loan repayments

Find an expression for the annual payment L for a loan of \$450 000 over 25 years at 6.5% p.a. compound interest.

Steps	Working
1 Write a recurrence expression for the amount A_n.	
2 Write expressions for the first few values.	
3 Simplify the expression for A_4.	
4 Use the geometric series formula $S_n = \dfrac{a(r^n - 1)}{r - 1}$	
5 Generalise the expression.	
6 The loan is paid off when the amount owing is 0.	
7 Write the equation.	
8 Solve for L.	
9 Write the answer.	

MATCHED EXAMPLE 29 | Population modelling

The population of youngsters in Queensland in 2022 was 2 500 000. Assuming an increase of 2% each year, write an expression for the likely population in 2030.

Steps	Working
1 Write a recurrence expression for the amount P_n.	
2 Write expressions for the first few years after 2022, where n is the number of years after 2022.	
3 Simplify the expression for P_3.	
4 Generalise the expression.	
5 Write an expression for 2030 ($n = 8$).	
6 Write the answer.	

PROOF AND NUMBER

MATCHED EXAMPLE 1	Show that a recurring decimal is a rational number

Express each recurring decimal in the fraction form.

a $1.\dot{4}$ **b** $0.5\dot{3}$ **c** $2.1\dot{2}5\dot{6}$

SB

p. 54

Steps	Working
a 1 Let x be the recurring decimal.	
2 Multiply both sides by 10 since there is only one recurring digit.	
3 Subtract equation [1] from equation [2].	
4 Solve the equation and write the answer.	

> You can check your answer using CAS by converting $1\dfrac{4}{9}$ to a decimal.

b 1 Let x be the recurring decimal.	
2 Multiply both sides by 100 since there are two recurring digits.	
3 Subtract equation [1] from equation [2].	
4 Solve the equation and write the answer.	
c 1 Let x be the recurring decimal.	
2 Multiply both sides by 10 so that the first digit after the decimal point is a recurring digit.	
3 Multiply both sides of equation [1] by 1000 since there are 3 recurring digits to eliminate.	

4 Subtract equation [1] from equation [2].

5 Solve the equation and write the answer.

SB

Using CAS 1:
Expressing
recurring decimals
in fraction form
p. 55

MATCHED EXAMPLE 2 | Describing number sets

SB

p. 55

State the smallest subset of R (N, Z, Q, R or C) that contains each number or set of numbers.

a $\sqrt{289}$

b $\{x : x^2 < 9\}$

c $27^{\frac{1}{3}} + 0.\dot{3}$

d $\{x : -7 < x < 5\}$

e $(2 + 3i) + (1 - 5i)$

Steps	Working
a 1 Simplify the expression, if possible. **2** Determine the smallest subset of C that contains 17.	
b 1 Simplify the inequality. **2** Determine the smallest subset of C that contains these values.	
c 1 Simplify, if possible. **2** Determine the smallest subset of C that contains the number obtained in step 1.	
d 1 Simplify, if possible. **2** Determine the smallest subset of C.	
e 1 Simplify the expression. **2** Determine the smallest subset of C.	

MATCHED EXAMPLE 3 Deciding if a set is closed

Determine if the set of rational numbers, Q, is closed under each of the following operations.

a subtraction   **b** multiplication   **c** division

Steps	Working
a 1 Show the operation in terms of the elements in the set of integers.	
2 State the conclusion.	
b 1 Show the operation in terms of the elements of the set.	
2 State the conclusion.	
c 1 Show the operation in terms of the elements of the set.	
2 State the conclusion.	

MATCHED EXAMPLE 4 | Use closure properties of a set to prove closure of another set

Given that the set of integers, Z, is closed under subtraction and multiplication, prove that the set of rational numbers, Q, is closed under subtraction.

Steps	Working
1 Represent x and y as a fraction.	
2 Subtract x from y.	
3 For $\dfrac{bc-ad}{db}$ to be rational, both $bc-ad$ and bd must be integers and $bd \neq 0$. Explain why $bc-ad$ and bd are integers and $bd \neq 0$.	
4 State the conclusion.	

MATCHED EXAMPLE 5 | Working with sets

Use the following sets to answer the questions:

$U = \{$integers from -2 to 12 inclusive$\}$

$A = \{$even integers from -1 to 5 inclusive$\}$

$B = \{$odd numbers from 0 to 9 inclusive$\}$

$C = \{$multiples of 4 from 3 to 12 inclusive$\}$

Find

a $A \cup B$ **b** $B \cap C$ **c** $\left(A \cap (B \cup C)\right)'$ **d** $n(A \cup C)'$

Steps	Working
a 1 List the elements of sets A and B.	
2 To find the union, list the elements in A and B, omitting duplicate elements.	
b 1 List the elements of sets C.	
2 Find the elements common to sets B and C.	
c 1 List the elements of $A \cap (B \cup C)$.	
2 Find the complement, which consists of all the elements in U that are not in $A \cap (B \cup C)$.	
d 1 Find the elements of $A \cup C$.	
2 List the elements of $(A \cup C)'$.	
3 State the number of elements in $(A \cup C)'$.	

MATCHED EXAMPLE 6 | Venn diagrams

Show each set as a Venn diagram.

a $U = \{$odd numbers between 2 and 23$\}$

$A = \{$odd numbers from 3 to 15 inclusive$\}$

b $U = \{$letters in the word *MATHEMATICS*$\}$

$B = \{$letters in the word *MATH*$\}$

$C = \{$letters in the word *TIMES*$\}$

Steps	Working
a 1 Only one circle is needed, and it contains all the elements given. Determine the elements of this set.	
2 Identify the elements in the universal set that are not in the given set. That is, find the complement of the set.	
3 Draw the Venn diagram.	
b 1 Decide if there is an intersection between the two subsets of the universal set.	
2 Find the elements remaining in each set that are not part of the intersection of the universal set.	
3 Find the elements remaining in the universal set.	
4 Draw the Venn diagram.	

MATCHED EXAMPLE 7 | Cardinal number with two sets

A survey of 40 students found that 15 like apples, 13 like oranges and 5 like both fruits. Display this information as a Venn diagram.

Steps	Working
1 Display the given information on a Venn diagram.	
2 Calculate all the other required information.	
3 Display the remaining information on the Venn diagram.	

MATCHED EXAMPLE 8 | Cardinal number with 3 sets

Use the following information to show a completed Venn diagram:

$n(U) = 120$

$n(A) = 46, n(B) = 35, n(C) = 47$

$n(A \cap B) = 17, n(A \cap C) = 30, n(B \cap C) = 22$

$n(A \cap B \cap C) = 8$

Steps	Working
1 Display the known information as a Venn diagram. Determine the values of some intersections that can be found immediately using the given information.	
2 Find the remaining values of the intersections and the number of elements outside the three sets.	
3 Draw the Venn diagram.	

SB

p. 66

MATCHED EXAMPLE 9 | Using a factor tree

a Construct a factor tree to express 120 as the product of prime numbers.

b State all the factors of 120 that are greater than 1.

Steps	Working
a 1 Find two factors whose product is 120.	
2 Repeat step 1 for any factor that is not a prime number. 2 is prime, but 60 is not and can be written as 2×30.	
3 Keep repeating step 1 until all the factors are prime numbers. 2 is prime, but 30 can be written as 5×6.	
4 Keep repeating step 1 until all the factors are prime numbers. 5 is prime, but 6 can be written as 2×3. Stop, because all the factors are prime.	
5 Write the number as the product of prime numbers (circled in the factor tree).	
b 1 The factors of 120 include all the prime factors and combinations of the products of the factors, including powers.	
2 State the factors.	

SB

p. 67

a Use a factor tree to find $\sqrt{620}$.

b Simplify $\sqrt{729}$.

Steps	Working
a **1** Construct a factor tree whose product of prime factors is 620.	
2 Simplify the surd using the powers of the prime numbers.	
b **1** Write 729 as the product of prime numbers.	
2 Evaluate the square root using the powers of the prime numbers.	

2

MATCHED EXAMPLE 11 | Proofs involving integers

a Prove that if m and n are both odd numbers, then $m + n$ is even.

b Prove that if m and n are both even numbers, then $m \times n$ is even.

Steps	Working
a 1 Write each number as an element of its set.	
2 Write an expression for $m + n$, and show that it is even.	
3 State the conclusion.	
b 1 Write each number as an element of its set.	
2 Write an expression for $m \times n$, and show that it is even.	
3 State the conclusion.	

9780170464109

SB

p. 68

a Given that the product of an odd and an even number is even, prove that $n^2 - n$ is even (where $n \in Z$).

b Hence, prove that if n^2 is odd, then n is odd.

Steps	Working
a **1** Write $n^2 - n$ as a product by factorising.	
2 Use the given property.	
3 State the conclusion.	
b Use the result from **a**.	

MATCHED EXAMPLE 13 | Proof by counterexample

Prove that each statement is false by providing a counterexample.

a All natural numbers are either prime or composite.

b If n is an integer, then n^2 is always even.

c If n is an even integer, then it is divisible by 4.

Steps	Working
a 1 Find a natural number that is neither prime nor composite.	
2 State the conclusion.	
b 1 Try values of n such that n^2 is odd.	
2 State the conclusion.	
c 1 Find an even integer that is not divisible by 4.	
2 State the conclusion.	

MATCHED EXAMPLE 14 | Applying direct proof

a Given that the sum of two even numbers is even, show that for even m, $m^2 + 6m$ is even.

b Show that if $n \in N$ is odd, then $2n + 3$ is odd.

Steps	Working
a 1 Factorise the expression to obtain the product of two quantities.	
2 Apply the given information.	
3 Write the complete expression to include the change.	
4 Draw the conclusion.	
b 1 Write the expression using the information provided.	
2 Write the result in a form from which the required conclusion can be made.	
3 State the conclusion.	

MATCHED EXAMPLE 15 | Applying proof by contradiction to surds

Prove by contradiction that the product of a rational and an irrational number is irrational.

Steps	Working
1 Assume that the statement is false.	
2 Rearrange the equation.	
4 State a conclusion.	
5 State the contradiction.	
6 State the conclusion.	

MATCHED EXAMPLE 16 | Proof by contradiction applied to prime numbers

Prove by contradiction that if $n > 1$ is not divisible by any prime number p, where $p \leq \sqrt{n}$, then n is a prime number.

Steps	Working
1 Assume that the statement is false.	
2 Write the given information.	
3 Create a number greater than p_n, and assume that it is composite.	
4 State the contradiction.	
5 State the conclusion.	

MATCHED EXAMPLE 17 | Proof by contradiction involving the set of integers

Prove by contradiction that for all $a, b \in Z$, if a and b are odd integers, then there does not exist an integer z such that $a^2 + b^2 = z^2$.

Steps	Working
1 Assume the statement is false.	
2 Apply the given information.	
3 Use the property.	
4 Substitute all values, and use algebra to simplify the equation.	
5 State the contradiction.	
6 State the conclusion.	

MATCHED EXAMPLE 18 | Proof by contrapositive

Given that all $n \in N \, \forall n \in \mathbb{N}$, prove that if $n^3 + 5$ is odd, then n is even.

Steps	Working
1 Express the statement 'if P, then Q' as 'if Q, then P'.	
2 Prove 'if not Q, then not P'.	
3 State the conclusion.	

MATCHED EXAMPLE 19 | Proof by induction of a series formula

Prove by induction that $100 + 95 + 90 + ... + (105 - 5n) = \dfrac{5n(41-n)}{2}, n \in \mathbb{N}$.

Steps	Working
1 Prove the base step.	
2 State the hypothesis and find the required expression.	
3 Write the function in the required form.	
4 State the conclusion.	

MATCHED EXAMPLE 20 The sum of the exterior angles of a polygon

Given that for $n \geq 3$ the sum of the interior angles of an n-sided polygon is $180(n - 2)$ degrees, prove by induction that the sum of its exterior angles is 360 degrees.

Steps	Working
1 Prove the base step.	
2 State the hypothesis and the required expression.	
3 Write the function in the required form.	
4 State the conclusion.	

SB

p. 79

MATCHED EXAMPLE 21 | Divisibility proof

Use proof by induction to show that for all natural numbers, $5^n - 1$ is divisible by 4.

Steps	Working
1 Verify the base step.	
2 State the hypothesis and the required expression.	
3 Write the function in the required form.	
4 State the conclusion.	

MATCHED EXAMPLE 22 | Inequality proof

Use proof by induction to show that $3^n > 3n$ for $n > 1$.

Steps	Working
1 Show the base step.	
2 State the hypothesis and the required expression.	
3 Write the function in the required form.	
4 State the conclusion.	

CHAPTER

3 GRAPH THEORY

MATCHED EXAMPLE 1 | Describing vertices and edges

Consider the graph shown.

a Which vertices are adjacent to vertex E?

b Which vertex has a loop?

c Which vertex is isolated?

d Which pair of vertices are connected by multiple edges?

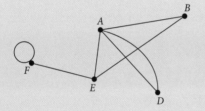

Steps	Working
a The vertices adjacent to vertex E are connected by an edge.	
b A loop connects a vertex to itself.	
c The isolated vertex is not connected to any other vertex.	
d Vertices connected by more than one edge have multiple edges.	

MATCHED EXAMPLE 2 | Using lists to represent a graph

Represent the graph using lists.

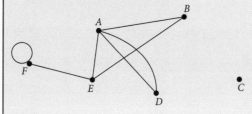

SB

p.92

Steps	Working
1 Write the list of vertices as a set.	
2 There are seven edges. Each edge is written using the letters of the vertices it connects. The loop at F is written as FF and the multiple edges between A and D are shown by writing AD twice.	

p. 93

MATCHED EXAMPLE 3 | Represent the graph using an adjacency matrix

Steps	Working
1 Draw a 4×4 matrix and label the rows and columns A, B, C and D.	
2 Row A There is no loop at A. Vertex A is connected to vertex B and C. Write 0 in cells AA and AD, and 1 in cells AB and AC.	
3 Complete the other rows in the same way.	

Cell AB is in row A and column B.
In an adjacency matrix, the value in
cell AB = the value in cell BA

MATCHED EXAMPLE 4 | Drawing a graph from lists

SB

p. 93

Draw the directed graph from the lists.

Vertices = {A, B, C, D, E}

Edges = {AA, AB, AC, AD, BC, BE, CD, DE}

Steps	Working
1 Vertices = {A, B, C, D, E} Draw the five vertices A, B, C, D and E.	
2 Edges = {AA, AB, AC, AD, BC, BE, CD, DE} Draw a loop at A for the edge AA. Draw a single edge between vertex A and vertex B for edge AB and repeat for edges AC, AD, BC, BE, CD and DE.	

MATCHED EXAMPLE 5   Drawing a map as a graph

The map shows the roads that connect towns A, B, C and D. Draw this map as a graph.

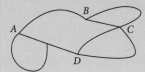

Steps	Working
1 Each of the four towns is a vertex. List the four vertices.	
2 Each possible path between the towns is an edge.	
From town A, there is one path to town B, one path back to town A and two paths to town D.	
From town B, there are two paths to town C.	
From town C, there are two paths to town D.	
3 Draw the four vertices.	
4 Draw all the edges.	
The graph contains a loop at vertex A.	

MATCHED EXAMPLE 6 | Finding the degree of the vertices of graph

Find the degree of each of the vertices in the graph shown.

Steps	Working
1 Three edges are connected to vertex *A*, so the degree is 3.	
2 Two edges are connected to vertex *B*, so the degree is 2.	
3 Only one edge is connected to vertex *C*, so the degree is 1.	
4 Two edges and a loop are connected to vertex *D*, so the degree is $2 + 2 = 4$.	

③

p. 96

MATCHED EXAMPLE 7 | Applying the handshaking lemma

A graph with 10 edges has 5 vertices A, B, C, D and E. The table below shows the degree of each vertex. Find the degree of vertex E.

Vertex	Degree
A	2
B	2
C	4
D	3
E	x

Steps	Working

1 Substitute $e = 10$ into the formula.

2 Add the degrees of the vertices in the table.

MATCHED EXAMPLE 8 | Classifying graphs

Classify the following graphs as simple, connected, complete, planar, bipartite or a combination of these.

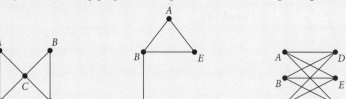

SB
p. 101

Steps	Working
a Simple: no loops or multiple edges. Connected: every vertex can be reached from every other vertex. Planar: no edges that cross.	
b Simple: no loops or multiple edges. Not connected: it is not possible to reach the lower right vertex from the other four vertices. Planar: The graph is not connected so it cannot be planar.	
c Simple: no loops or multiple edges. Connected: every vertex can be reached from every other vertex. Bipartite: there are two groups of vertices with edges connecting the vertices in one group to the vertices in the other group. Non-planar: it cannot be redrawn so that no edges cross.	

MATCHED EXAMPLE 9 | Finding the number of edges in a complete graph

How many edges would there be in a complete graph with eight vertices?

Steps	Working
A complete graph with n vertices has $\dfrac{n(n-1)}{2}$ edges. Substitute $n = 8$ into the formula.	

MATCHED EXAMPLE 10 | Vertices and edges in a complete graph

Determine the number of vertices and edges for the complete graph K_{10}.

Steps	Working
The complete graph K_n has n vertices and $\dfrac{n(n-1)}{2}$ edges. Substitute $n = 10$ into the formula.	

p. 102

MATCHED EXAMPLE 11 | Subgraphs

Which of the following graphs is not a subgraph of the graph shown?

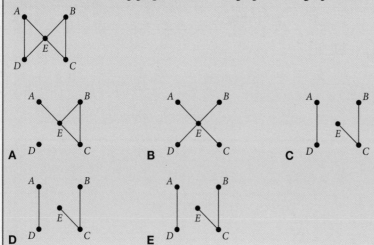

Steps	Working
The original graph has	

vertices = {A, B, C, D, E}

edges = {AD, AE, BC, BE, CE, DE}

A subgraph must also contain vertices and edges
chosen from these sets.

MATCHED EXAMPLE 12	Finding the number of edges in a regular graph	

Determine the number of edges for a regular graph with five vertices of degree 4.

Steps	Working
A regular graph with n vertices each of degree r has $\dfrac{nr}{2}$ edges. Substitute $n = 5$ and $r = 4$ into the formula.	

MATCHED EXAMPLE 13 | Identifying the complement of a graph

The complement of graph G is

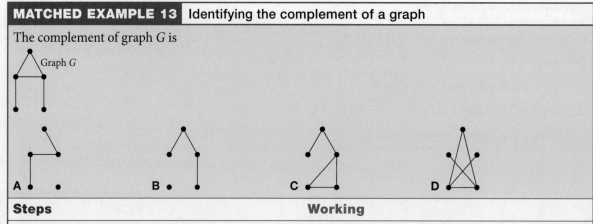

Steps	Working
The complete graph K_5 has 10 edges.	
The complement of G contains the other five edges that are in the complete graph K_5 but are not present in G.	

MATCHED EXAMPLE 14 | Identifying planar graphs

Which of the connected graphs below are planar?

a

b

Steps	Working
a The graph can be redrawn so that no edges cross.	
b The graph cannot be drawn so that no edges cross. This is the complete graph K_5, so it cannot be planar.	

p. 107

MATCHED EXAMPLE 15 Using Euler's rule with an adjacency matrix

The adjacency matrix represents a planar graph.

$$\begin{array}{c} \\ A \\ B \\ C \\ D \end{array} \begin{array}{c} \begin{array}{cccc} A & B & C & D \end{array} \\ \begin{bmatrix} 0 & 1 & 0 & 1 \\ 1 & 0 & 1 & 0 \\ 0 & 1 & 0 & 1 \\ 1 & 0 & 1 & 0 \end{bmatrix} \end{array}$$

Determine the number of faces.

Steps	Working
1 The number of edges is 4, found by adding the numbers above or on the leading diagonal of the matrix. The graph has four vertices $\{A, B, C, D\}$ and four edges $\{AB, AD, BC, CD\}$.	
2 Substitute into Euler's rule to determine the number of faces.	

9780170464109

MATCHED EXAMPLE 16 | Applying Euler's rule

A connected planar graph has seven vertices and nine edges

Determine the number of faces.

Steps	Working
Use Euler's rule to determine the number of faces. $v + f - e = 2$	

3

p. 110

MATCHED EXAMPLE 17 | Interpreting a bipartite graph

The bipartite graph shows the tasks the four workers are qualified to perform.

Determine which of the following is a valid allocation.

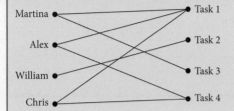

A

Martina	Task 1
Alex	Task 3
William	Task 2
Chris	Task 4

B

Martina	Task 3
Alex	Task 4
William	Task 2
Chris	Task 1

C

Martina	Task 2
Alex	Task 3
William	Task 1
Chris	Task 4

D

Martina	Task 3
Alex	Task 2
William	Task 4
Chris	Task 1

E

Martina	Task 1
Alex	Task 3
William	Task 2
Chris	Task 4

Steps	Working
There must be an edge between the worker and the task for the allocation to be valid. Each option must be checked on the bipartite graph to see if there is an edge between the worker and the task number.	

MATCHED EXAMPLE 18 | Maximum edges for a complete bipartite graph

Find the maximum number of edges for a complete bipartite graph with four vertices.

Steps	Working
1 Determine the different possible values of m and n in the complete bipartite graph $K_{m,n}$.	
2 The complete bipartite graph $K_{m,n}$ has $m \times n$ edges.	

p. 112

MATCHED EXAMPLE 19 | Identifying trees

Which of the following graphs are trees?

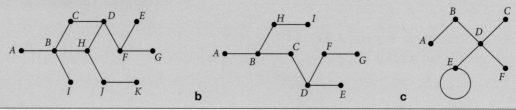

a **b** **c**

Steps	Working
a This graph is not a tree as it contains a cycle *CDHBC*.	
b This graph is a tree as it is a simple connected graph with no cycles.	
c This graph is not a tree. It is not a simple graph as it contains a loop at vertex *E*.	

   9780170464109

MATCHED EXAMPLE 20	Finding the number of vertices in a tree

SB

p. 112

Find the number of vertices in a tree with seven edges.

Steps	Working
A tree with n vertices contains $n - 1$ edges.	
Substitute $e = 7$ into the formula $e = v - 1$.	

MATCHED EXAMPLE 21 | Classifying travelling

For the graph shown, classify the following walks as either an open walk, a closed walk, a trail, a path, a circuit, a cycle or a combination of these.

a $A–B–C–E–D–B$

b $E–B–D–E$

c $A–B–C–E–D$

d $D–B–A–E–D$

Steps	Working
a A trail has no repeated edges and starts and finishes at different vertices.	
b A cycle is a path that starts and finishes at the same vertex.	
c A trail has no repeated edges and starts and finishes at different vertices. A path is a trail where there are no repeated vertices.	
d A cycle is a path that starts and finishes at the same vertex.	

9780170464109

MATCHED EXAMPLE 22 | Finding Eulerian trails

An Eulerian trail starting at vertex *A* exists for the following graph.

a Find the finishing vertex.

b Find the possible Eulerian trails.

Steps	Working
a 1 Find the degree of each vertex.	
2 The Eulerian trail must start and finish at the vertices with odd degree.	
b Trace all the possible paths that start at *B*, finish at *A* and include every edge in the graph.	

MATCHED EXAMPLE 23 | Finding Hamiltonian paths and cycles

The start of a Hamiltonian path is shown in blue.

a Complete the Hamiltonian path.

b Complete the Hamiltonian cycle.

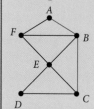

Steps	Working
a The Hamiltonian path must visit each vertex in the graph exactly once and finish at a different vertex to A.	
b The Hamiltonian cycle must visit each vertex in the graph exactly once and finish at the vertex A.	

LOGIC AND ALGORITHMS

MATCHED EXAMPLE 1	Understanding statements, premises and conclusions

SB

p. 131

The following is a logical argument involving Amanda and Ellie.

Amanda:	I have a cat.	1
Ellie:	Oh great. Can you bring it home tomorrow?	2
Amanda:	No, I can't.	3
Ellie:	But why?	4
Amanda:	Cats with long hair have lots of flea.	5
Ellie:	So what?	6
Amanda:	Cats with long hair shed all over the house.	7
Ellie:	Don't get a cat with long hair.	8

a Name the statements from which no conclusion is drawn.

b Name the premises and conclusion(s).

c Identify the sentences that are not statements.

Steps	Working
a Statements are either true or false and do not lead to a conclusion being made.	
b A premise is a statement from which a conclusion can be drawn.	
c Identify the sentences that cannot be considered to be true or false.	

MATCHED EXAMPLE 2 | Identifying logical connectives

Let A be 'Zuri is a singer' B be 'Karla is a rapper'

         C be 'Diana is a dancer' D be 'Ella is an actor'

Write each compound proposition using correct grammar.

a $A \wedge D$ **b** $D \rightarrow \neg C$ **c** $\neg A \rightarrow B$ **d** $\neg D \rightarrow (C \vee B)$

Steps	Working
a Write the conjunction in statement form.	
b **1** Write in statement form.	
2 Write using correct grammar.	
c **1** Explain each part of the compound proposition.	
2 Write in statement form.	
3 Write using correct grammar.	
d **1** Write each part of the conditional statement using proper grammar.	
2 Write the entire proposition using correct grammar.	

MATCHED EXAMPLE 3 | Truth table involving conjunction

Construct a truth table for $\neg A \wedge B$.

Steps	Working
1 Set up the main structure of the truth table.	
2 Add a column for the value of each combination of T and F.	

p. 136

MATCHED EXAMPLE 4 Truth table involving NOT and disjunction

Construct a truth table to show that $\neg(\neg A \vee B) \equiv A \wedge \neg B$.

Steps	Working
1 Set up the main structure of the truth table.	
2 Complete columns 1 and 2 with all possible combinations of T and F for A and B. Then fill in columns 3 and 4 by negating (taking the opposite) of each truth value in columns 1 and 2.	
3 Complete the columns for $\neg A \vee B$ and $\neg(\neg A \vee B)$.	
4 Complete the last column $A \wedge \neg B$.	
5 State the conclusion.	

9780170464109

Construct a truth table for the conditional statement

'If I eat ice cream, I will gain weight.'

p. 137

Steps	Working
1 Write the statements required for the truth table.	
2 Decide what each column of the truth table represents.	
3 Determine the result of each pair of truth values.	
4 Summarise the results as a truth table.	

4

MATCHED EXAMPLE 6 | Proving tautologies

Show with a truth table that $(A \rightarrow B) \vee (B \rightarrow A)$ is a tautology.

Steps	Working
1 Write the statements required for the truth table. Complete columns 1 and 2 with the four combinations of T and F.	
2 Complete the table by working out the truth value of each pair of truth values. $A \vee B$ is true if either A or B is true. $(A \rightarrow B) \vee (B \rightarrow A)$ is false only when $(A \rightarrow B)$ is false and $(B \rightarrow A)$ is false (see Worked example 5).	
3 State the conclusion.	

Show with a truth table the statement $B \land \neg(A \rightarrow B)$ is a contradiction.

Steps	Working
1 Write the statements required for the truth table. Complete columns 1 to 4, remembering that $A \rightarrow B$ is false when A is true and B is false. Complete the values for $\neg(A \rightarrow B)$.	
2 $B \land \neg(A \rightarrow B)$ is true if B is true and if $\neg(A \rightarrow B)$ is true. This is never the case, so $B \land \neg (A \rightarrow B)$ is false in all cases.	
3 State the conclusion.	

MATCHED EXAMPLE 8 | Change between base 2 and base 10

a Change 47 to base 2.   **b** Express 100110 in base 10 notation.

Steps	Working
a 1 Divide the dividend by 2 and write the remainder.	
2 Divide the new dividend by 2 and write the remainder.	
3 Repeat the method of division by 2 until the dividend is 0.	
4 Write the digits in reverse order.	
5 State the answer.	
b 1 Use the digits as the coefficients of powers of 2.	
2 State the answer.	

MATCHED EXAMPLE 9	Addition and subtraction using base 2

p. 143

Express the answer to each of the following in base 2.

a $100_2 + 110_2$ **b** $1010_2 - 1001_2$

Steps	Working
a 1 Set out the addition in the required layout.	
2 Perform addition, ensuring that each sum digit is binary.	
3 State the answer.	
b Set out the addition in the required layout.	
2 Perform the subtractions, converting digits to binary form if necessary.	
3 State the answer.	

MATCHED EXAMPLE 10 | Multiplication using base 2

a Perform the multiplication $101_2 \times 10_2$ **b** Write $101_2 \times 10_2$ and the answer using base 10.

Steps	Working
a 1 Set out the multiplication in the required form.	
2 Perform the multiplications.	
3 Perform the addition.	
4 State the answer.	
b 1 Write each binary number in base 10.	
2 Write the binary answer in base 10 form.	
3 State the answer.	

MATCHED EXAMPLE 11	Division using base 2

Carry out the division $11100_2 \div 10_2$

Steps	Working

1 Display the divisor and the dividend in the form of long division.

2 Starting from the left, see if the divisor divides the first digit of the dividend.

3 Multiply by the dividend, subtract and bring down the next digit.

4 Check to see if the divisor divides the result.

5 Multiply by the dividend, subtract and bring down the next digit.

6 Check to see if the divisor divides the result.

7 Multiply by the dividend, subtract and bring down the next digit.

8 Check to see if the divisor divides the result.

9 Multiply by the dividend, subtract and bring down the next digit.

10 Check to see if the divisor divides the result.

11 Multiply by the dividend, subtract and bring down the next digit.

Stop when the subtraction is 0.

12 State the result.

MATCHED EXAMPLE 12 | Converting between decimal and binary fractions

a Write 1010.011_2 as a base 10 number.

b Express 12.25 in binary form.

c State 0.6 in binary form.

Steps	Working
a 1 Convert the whole part to a decimal.	
2 Convert the fraction part to a decimal.	
3 State the result.	
b 1 Convert the whole number part to a decimal.	
2 Convert the fraction part to a decimal.	
3 State the answer.	
c 1 Convert the fraction part to a decimal.	
3 State the answer.	

MATCHED EXAMPLE 13 | Proof using Boolean algebra

Use Boolean algebra to show that $(X + Y)(\overline{X} + Y) = Y$.

Steps	Working
1 Apply the distributive law.	
2 Simplify by applying the appropriate Boolean laws.	

MATCHED EXAMPLE 14 | Applying Boolean laws to the distributive law

Simplify $(X\bar{Z} + \bar{Y}X) \bullet (X\bar{Z} + \bar{X})$.

Steps	Working
1 Apply the distributive law.	
2 Simplify by applying the appropriate Boolean laws.	
3 State the answer.	

4

MATCHED EXAMPLE 15 | Sum of products for a 3-variable truth table

The truth table is shown for three variables. Show that the SOP is $C\overline{A} + CB$.

A	B	C	Output
0	0	0	0
0	0	1	1
0	1	0	0
1	0	0	0
1	0	1	0
0	1	1	1
1	1	0	0
1	1	1	1

Steps | **Working**

1 Write the AND expression (product) for the output in each row that is 1.

2 State the sum of products, SOP.

3 Minimise the product sum.

MATCHED EXAMPLE 16 | Finding the sum of products

Use a truth table to find the SOP for $(A + \overline{B})(A + C)$.

Steps	Working
1 Complete the truth table.	
2 Write the product for each row that is not 0.	
3 State the sum of products, SOP.	
4 Simplify the product sum.	

p. 154

MATCHED EXAMPLE 17 | Application of Boolean operators

Write Boolean instructions that will perform each of the following searches.

a Red or green in marbles

b Sports, excluding football

c Singing and dancing, but excluding classical and salsa

Steps	Working
a 1 There are three keywords. Decide which Boolean operators to use. **2** Write the Boolean search query.	
b 1 There are two keywords/phrases. Decide which Boolean operators to use. **2** Write the Boolean search query.	
c 1 There are four keywords/phrases. Decide which Boolean operators to use. **2** Write the Boolean search query.	

9780170464109

MATCHED EXAMPLE 18 | Applying the conditional statement

SB

p. 156

A conditional statement takes the form if (logical test, value if true, value if false).

The negation, NOT, is of the form NOT(logical test).

Complete the table according to the following.

a In column C, each cell has '1' if corresponding entries in columns A and B are the same and '0' if cells in columns A and B are different.

b In column D, each cell is the negation of its corresponding cell in column C.

	A	B	C	D
1	Cairo	Cairo		
2	Madrid	Athens		
3	Paris	Paris		
4	Berlin	Rome		
5	Santiago	Santiago's		

Steps	**Working**
a **1** Determine the 'logical test', 'value if true' and 'value if false' in if (logical test, value if true, value if false).	
2 Write the statement for if (logical test, value if true, value if false).	
b **1** Use NOT(logical test) applied to column C.	

SB

p. 161

MATCHED EXAMPLE 19 | Using a Karnaugh map to minimise SOP

Use the K-map to find the minimised expression for the sum of the products.

CD AB	00	01	11	10
00	1	1	1	1
01	1	0	0	1
11	0	1	1	0
10	1	1	1	0

Steps | **Working**

1 Identify the groups to use.

2 Write the expression for each group.

4 Write the expression for the SOP.

MATCHED EXAMPLE 20 | Constructing a Karnaugh map

Construct a K-map from the truth table and find the minimised expression for the SOP.

A	B	C	D	Product
0	0	0	0	1
0	0	0	1	1
0	0	1	0	1
0	0	1	1	1
0	1	0	0	1
0	1	0	1	0
0	1	1	0	1
0	1	1	1	1
1	0	0	0	1
1	0	0	1	1
1	0	1	0	1
1	0	1	1	1
1	1	0	0	1
1	1	0	1	0
1	1	1	0	1
1	1	1	1	0

Steps	Working
1 Complete the K-map.	
2 Identify the groups and determine the product terms.	
3 Write the expression for the SOP.	

MATCHED EXAMPLE 21 Determine the sum of products from a truth table

a Find the SOP for the truth table shown.

b Deduce what the SOP will be if the negated value of each cell in the truth table is used.

A	B	C	Product
0	0	0	0
0	0	1	0
0	1	0	1
0	1	1	1
1	0	0	0
1	0	1	0
1	1	0	1
1	1	1	1

Steps	Working
a 1 Complete the K-map.	
2 Identify the groups and calculate the product terms. In the only group, $B = 1$ does do not change. This is B.	
3 Find the SOP.	
b Write the negation.	

MATCHED EXAMPLE 22 | Truth table for an OR gate

Construct the truth table for the logic gate shown.

A
B — Z
C

Steps	Working
1 Identify the logic gate required.	
2 Construct the truth table.	

MATCHED EXAMPLE 23 | Find the output function from a circuit

Determine the output function, Z, for the circuit shown.

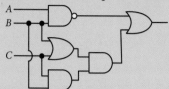

Steps	Working
1 Start from the left and identify the first logic gate.	
2 Describe the next gate.	
3 Describe the remainder of the circuit.	
4 State the output for the circuit.	

MATCHED EXAMPLE 24	Constructing a logic gate using NOT, OR and AND

Show $\overline{A}(B + C)$ as a logic gate diagram.

Steps	Working
1 Identify the logic gates required.	
2 Join the gates in the correct order.	

MATCHED EXAMPLE 25	Logic gate using OR and AND

Show $(A + B)(C + D)$ as a logic gate diagram.

Steps	Working
1 Working left to right, identify the first section and gate.	
2 Identify the next section and gate.	
3 Complete the circuit diagram by combining the main sections.	

MATCHED EXAMPLE 26 | Simplifying a circuit

Draw a simplified version of the circuit shown below.

p. 170

Steps	Working
1 Working from left to right, describe each gate.	
2 Simplify the final output using any method.	
3 Draw the simplified circuit.	

SB

p. 175

MATCHED EXAMPLE 27 | Designing a flowchart

Draw a flowchart to show the following steps in the algorithm to purchase a shirt.

- Pick a shirt.
- Check if the size is OK.
- If the size is fine, check if it fits well.
- If it fits well, buy it.
- If it doesn't fit, search for another shirt and repeat the method above.
- If the size isn't OK, take another shirt and repeat the method above.

Steps	Working
1 Use a start/end symbol and perform the first instruction.	
2 Show the decisions to be made after a shirt is picked.	

3 Complete the flowchart.

p. 177

MATCHED EXAMPLE 28 | Interpret a flowchart

The flowchart below describes an algorithm.

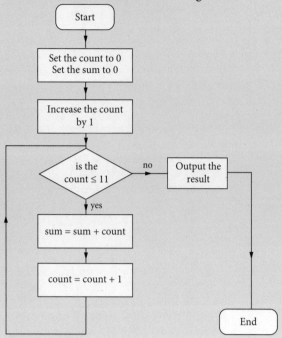

a State the values for **count** and **sum**.

b Describe what the algorithm is calculating.

Steps	Working

a 1 State the initial values.

Apply the flowchart instructions for the first value of count.

2 Follow the correct path for the decision.

3 Repeat the loop until the count value is greater than 35.

9780170464109

4 Follow the final set of instructions prior to terminating the program.

b **1** Describe the situation by looking at what happens in the process symbols.

2 State the conclusion.

MATCHED EXAMPLE 29 | Interpreting pseudocode

a Describe what the program below is doing.

INTEGER array[20]	line 1
count←0	line 2
FOR loop = 1 TO 20	line 3
IF loop MOD 5 = 3	line 4
THEN	line 5
count←count + 1	line 6
array[count] ← loop	line 7
ENDIF	line 8
ENDFOR	line 9
PRINT "Numbers from 1 to 20 with remainder 3 after division by 5 are"	line 10
FOR numeral = 1 TO count	line 11
PRINT array[numeral]	line 12
ENDFOR	line 13

Steps	Working
a Describe what is happening in each section.	
b **1** Show the output from the first PRINT statement.	
2 Write the output from the second PRINT command.	

MATCHED EXAMPLE 30	Pseudocode involving addition of numbers	

SB

p. 180

Write pseudocode that finds the sum of 30 random numbers till 50 and prints the result.

Steps	Working
1 Set an array to store 30 numbers and initialise the sum of the numbers and set up a loop to read the numbers.	
2 Choose a random number.	
3 The procedure keeps a running total of the numbers.	
4 Print the result.	

MATCHED EXAMPLE 31 | Pseudocode using random numbers, modulus, product

The function A MOD B returns the remainder after A is divided by B, for example, 8 MOD 3 = 2.

The function RAND(A) generates a random whole number from 1 to A, for example, RAND(5) = 3.

Write pseudocode that generates six random even whole numbers from 2 to 34 and then finds their sum.

Steps	Working
1 Initialise the product of the numbers and create a loop for the random numbers.	
2 Check to see if the random number is even. Print the even number.	
3 Update the sum and output the result.	

MATCHED EXAMPLE 32 | Integer array

Use pseudocode to store the even numbers 1 to 50 in reverse order.

Steps	Working
1 Store the numbers as an array.	
2 Create a second array and store the even numbers in reverse order.	

CHAPTER

5 MATRICES

SB

p. 197

MATCHED EXAMPLE 1	Matrix dimensions and elements

$P = \begin{bmatrix} 5 & -6 & 4 \\ 2 & 8 & -1 \end{bmatrix}$, $Q = \begin{bmatrix} 5 & -7 & 8 \end{bmatrix}$ and $R = \begin{bmatrix} -2 & 4 \\ 3 & 4 \end{bmatrix}$

a What are the dimensions of each matrix above?

b What are p_{23}, p_{12}, q_3 and r_{21}?

Steps	Working
a Write the number of rows and columns.	
b Select the correct rows and columns.	

MATCHED EXAMPLE 2 | Basic operations

$P = \begin{bmatrix} 4 & -3 & 2 \\ 5 & 1 & -7 \end{bmatrix}$, $Q = \begin{bmatrix} -10 & -2 & 0 \\ 12 & 5 & 8 \end{bmatrix}$ and $R = \begin{bmatrix} 0 & 8 & -7 \\ 8 & -1 & 11 \end{bmatrix}$

Find $3P$, $-Q$, $P + R$, $Q + (-Q)$, $Q - R$ and $4P + 5R$.

Steps	Working
1 Multiply every element by 3.	
2 Change the sign of every element.	
3 Add corresponding elements.	
4 Add the matrices. $Q + (-Q) = 0$, no matter what the particular matrix Q happens to be.	
5 Subtract corresponding elements.	
6 Do the multiplications first, then do the addition.	

MATCHED EXAMPLE 3 | Matrix multiplication

$P = \begin{bmatrix} 3 & 1 \\ -1 & 0 \\ 2 & 2 \end{bmatrix}$ and $Q = \begin{bmatrix} -3 & 4 & 8 \\ 1 & 5 & -6 \end{bmatrix}$

Find PQ and QP.

Steps	Working
1 Write the product PQ.	
2 Each element is done as *row* × *column*.	
3 Write the product QP.	
4 Each element is done as *row* × *column*.	
5 PQ and QP are not the same.	

SB
p. 200

$A = \begin{bmatrix} 0 & 1 \\ 4 & 7 \end{bmatrix}$ $B = \begin{bmatrix} 3 \\ 5 \end{bmatrix}$ and $C = \begin{bmatrix} 2 \\ -5 \end{bmatrix}$ Find

a $A(B+C)$ **b** $AC + B$

Steps	Working
a 1 Perform the operation in brackets first.	
2 Perform the remaining operation and simplify.	
b 1 Perform multiplication before addition.	
2 Perform the remaining operation and simplify.	

SB
Using CAS 1:
Matrix operations
p. 201

MATCHED EXAMPLE 5 | Matrix simplification

Simplify each of the following.

a $5A - 8B + 9A + 3B$

b $5(C - 8F) - 3(3C + 6F)$

c $2XY + 5YX - 8YX + 7XY$

d $7PQ + 8P - 6Q + 4PQ - 2P + 2QP - 7Q$

Steps	Working
a **1** Rearrange using commutativity.	
2 Use the distributive law and simplify.	
b **1** Use the distributive law.	
2 Use commutativity and the distributive law.	
3 Simplify.	
c **1** Rearrange.	
2 Use the distributive law and simplify.	
d **1** Rearrange.	
2 Use the distributive law and simplify.	

9780170464109

MATCHED EXAMPLE 6	Expansion of matrix expressions

Expand the brackets of the following.

a $M(3N + 5L)$ **b** $4A(7B - 5I)$ **c** $(3X - 4Y)(X - 2Y)$

Steps	Working
a **1** Use the distributive law.	
2 Use the associative law.	
b **1** Use the distributive law.	
2 Use the associative law (twice) and simplify.	
3 Use the identity law $AI = A$.	
c **1** Use the distributive law.	
2 And again.	
3 Simplify.	
4 Write $XX = X^2$.	

5

MATCHED EXAMPLE 7	Factorisation of matrix expressions

Factorise the following as much as possible.

a $2XYZ - 4XTZ$ **b** $7AB + 2B$ **c** $8P^2 - Q^2$ **d** $16A^2 + 20AB - 20BA - 25B^2$

Steps			Working
a	**1**	Use the distributive law for scalar multiplication.	
	2	Use associativity.	
	3	Use the left distributive law.	
	4	Use the right distributive law.	
b	**1**	Use $B = IB$ to insert the identity on the correct side.	
	2	Use the right distributive law.	
c	**1**	Insert $2PQ - 2PQ$.	
	2	Use the left and right distributive laws.	
d	**1**	Use the left distributive law.	
	2	Use the right distributive law.	

MATCHED EXAMPLE 8 | Inverse of a diagonal matrix

p. 209

Find the inverse of $M = \begin{bmatrix} 7 & 0 \\ 0 & 4 \end{bmatrix}$, if it exists.

Steps	Working
1 Choose variables for a 2×2 matrix.	
2 Find the product of M and X.	
3 Write $MX = I$.	
4 Solve for a, b, c, d.	
5 Write the answer.	

MATCHED EXAMPLE 9 Determinant of a diagonal matrix

Use the determinants to find the nature of the following.

a $A = \begin{bmatrix} -2 & 0 \\ 0 & 5 \end{bmatrix}$ **b** $B = \begin{bmatrix} 8 & 0 & 0 \\ 0 & 0 & 0 \\ 0 & 0 & 4 \end{bmatrix}$ **c** $C = \begin{bmatrix} 1 & 0 & 0 & 0 \\ 0 & -5 & 0 & 0 \\ 0 & 0 & 2 & 0 \\ 0 & 0 & 0 & 6 \end{bmatrix}$

Steps	Working
a **1** State the reason and multiply the diagonal elements to find the determinant.	
2 Sate the nature of A.	
b **1** State the reason and multiply the diagonal elements to find the determinant.	
2 State the nature of B.	
c **1** State the reason and multiply the diagonal elements to find the determinant.	
2 State the nature of C.	

Find the inverses of $P = \begin{bmatrix} 4 & -3 \\ 0 & 1 \end{bmatrix}$ and $Q = \begin{bmatrix} 5 & -3 & 1 \\ 0 & 2 & 4 \\ 0 & 0 & 0 \end{bmatrix}$, if they exist.

Steps	Working
a 1 Choose variables for a 2×2 matrix.	
2 Find PX.	
3 Write $PX = I$.	
4 Solve.	
5 Write the answer.	
b 1 Choose variables for a 3×3 matrix.	
2 Find QX.	
3 Write $QX = I$.	
4 Solve for a, b, c, etc.	
6 Write the result.	

MATCHED EXAMPLE 11 | Determinant of a triangular matrix

a $A = \begin{bmatrix} -3 & 2 \\ 0 & 1 \end{bmatrix}$ **b** $B = \begin{bmatrix} 2 & 0 & 0 \\ 8 & 1 & 0 \\ -1 & 2 & 7 \end{bmatrix}$ **c** $C = \begin{bmatrix} 2 & -1 & 6 & 4 \\ 0 & -3 & -4 & -9 \\ 0 & 0 & 5 & 3 \\ 0 & 0 & 0 & 6 \end{bmatrix}$

Steps	Working
a 1 State the reason and multiply the diagonal elements to find the determinant.	
2 State the nature of A.	
b 1 State the reason and multiply the diagonal elements to find the determinant.	
2 State the nature of B.	
c 1 State the reason and multiply the diagonal elements to find the determinant.	
2 State the nature of C.	

Find the determinant and inverse of each matrix, if they exist.

a $A = \begin{bmatrix} 1 & 3 \\ 6 & 5 \end{bmatrix}$ **b** $\begin{bmatrix} 2 & -4 \\ 1 & -2 \end{bmatrix}$ **c** $C = \begin{bmatrix} 3 & 0 \\ -4 & 5 \end{bmatrix}$

SB p. 212

Steps	Working
a 1 Use $ad - bc$ for the determinant.	
2 State the outcome.	
3 Find the inverse.	
4 Write the answer.	
b 1 Use $ad - bc$ for the determinant.	
2 State the result.	
c 1 Use $ad - bc$ for the determinant.	
2 $|C| \neq 0$, so find the inverse.	
3 Write the answer.	

MATCHED EXAMPLE 13 | Determinant of a 3 × 3 matrix

Find the determinant of each matrix.

a $A = \begin{bmatrix} 5 & 0 & 2 \\ 1 & 3 & 4 \\ -1 & 1 & 0 \end{bmatrix}$

b $B = \begin{bmatrix} 2 & 1 & 3 \\ 4 & -2 & -1 \\ -5 & 2 & 6 \end{bmatrix}$

Steps	Working
a 1 Choose the row or column with the easiest numbers.	
2 Use the formula to expand along row 3.	
3 Find A_{31}: miss out the third row and the first column.	
4 Find A_{32}: miss out the third row and second column.	
5 Find $\det(A)$ using the expansion along row 3.	
b 1 Choose the row or column with the easiest numbers.	
2 Use the formula to expand along column 2.	
3 Find B_{12}: miss out the first row and second column.	
4 Find B_{22}: miss out the second row and second column.	
5 Find B_{32}: miss out the third row and second column.	
6 Find $\det(B)$ using the expansion along column 2.	

Find the determinants of $G = \begin{bmatrix} 3 & 1 & 2 \\ -1 & -2 & 5 \\ 2 & 4 & 1 \end{bmatrix}$

Steps	Working
1 Write the elements twice and put in the diagonal lines.	
2 Multiply and add forwards.	
3 Multiply and add backwards.	
4 Subtract the answers.	

Using CAS 2:
Determinants
p. 215

MATCHED EXAMPLE 15 | Simple matrix equations

Find these matrix equations.

a $5X - \begin{bmatrix} 2 & 4 \\ 3 & 7 \end{bmatrix} = \begin{bmatrix} 1 & 3 \\ 2 & 0 \end{bmatrix}$ **b** $4Y + \begin{bmatrix} 2 & 3 & 1 \\ 0 & 2 & 6 \end{bmatrix} = 5\left(\begin{bmatrix} 6 & 1 & 2 \\ 5 & 3 & 4 \end{bmatrix} - Y \right)$

Steps	Working
a 1 Add $\begin{bmatrix} 2 & 4 \\ 3 & 7 \end{bmatrix}$ to both sides and simplify.	
2 Multiply both sides by $\frac{1}{5}$ and simplify. Division of a matrix is not defined, so multiply by the reciprocal.	
b 1 Multiply out the brackets and simplify.	
2 Isolate Y and simplify.	
3 Multiply both sides by $\frac{1}{9}$ and simplify.	

MATCHED EXAMPLE 16 | Solution of matrix equations involving 2 × 2 matrices

Solve each matrix equation.

a $\begin{bmatrix} 2 & 4 \\ 1 & 3 \end{bmatrix} X = \begin{bmatrix} 1 & 2 \\ 3 & 0 \end{bmatrix}$
   b $X \begin{bmatrix} 5 & 0 \\ 2 & 1 \end{bmatrix} = \begin{bmatrix} 1 & 4 \\ -2 & 6 \end{bmatrix}$

c $\begin{bmatrix} 3 & 0 \\ 1 & -2 \end{bmatrix} X - \begin{bmatrix} 1 & -3 \\ -7 & 2 \end{bmatrix} = 4X + \begin{bmatrix} 3 & 7 \\ 4 & 2 \end{bmatrix}$

Steps	Working
a 1 Find the inverse of the coefficient matrix.	
2 Multiply both sides of the equation by the inverse *on the left side* to find the identity.	
3 Simplify and write the answer.	
Actually, do the multiplication that should give *I* to check the inverse is correct.	
b 1 Find the inverse of the coefficient matrix.	
2 Multiply both sides of the equation by the inverse on the *right* side.	
3 Simplify and write the answer.	
c 1 Move variables to the LHS and constant matrices to the RHS.	
2 Insert *I* on the same side of *X* as the other coefficient matrix.	

3 Factorise and simplify.

4 Find the inverse of the coefficient matrix.

5 Multiply both sides by the inverse on the left side.

6 Simplify and write the answer.

Solve $\begin{bmatrix} 6 & 1 \\ 2 & -4 \end{bmatrix} X = \begin{bmatrix} 1 \\ 3 \end{bmatrix}$.

SB
p. 221

Steps	Working
1 Find the inverse of the coefficient matrix.	
2 Left multiply both sides by the inverse.	
3 Simplify and write the answer.	

5

SB

Using CAS 4:
Matrix equations
p. 222

p. 234

MATCHED EXAMPLE 1	Applying the addition principle

For dinner, Daniel can choose only one item from: 4 different types of steak; or ravioli or spaghetti or lasagna; or grilled or fried chicken. How many choices does Daniel have for dinner?

Steps	**Working**
Daniel can choose only one item and in each case the word OR can be used between each of the choices. The word OR between the choices indicates the addition principle is used.	
Calculate the possibilities by adding the different options.	

MATCHED EXAMPLE 2 | Applying the multiplication principle

Alina must cook one main course and one dessert for dinner. She is told to prepare the main course using one vegetable out of six different vegetables and the dessert using one fruit out of three different fruits. How many different possibilities does she have?

Steps	Working
Alina must choose one vegetable from six vegetables AND one fruit from three fruits. AND indicates that the multiplication principle is used.	

6

MATCHED EXAMPLE 3 Using the addition and multiplication principles

Jessie is ordering lunch at a restaurant. She will order a pancake or noodles. The menu has pancakes with a choice of 2 different batters and 8 different syrups. The menu also has a choice of 3 different noodles with 4 different sauces. How many choices does Jessie have?

Steps	Working
1 Find the number of pancake choices.	
There are 2 different batters AND 8 different syrups.	
AND is the multiplication rule.	
2 Find the number of noodles choices.	
There are 3 different noodles AND 4 different sauces.	
3 Find the total number of choices.	
Jessie can have pancake OR noodles.	
OR is the addition rule.	

MATCHED EXAMPLE 4 | Applying tree diagrams

A box contains 2 red marbles, 1 blue marble and 4 green marbles. Two marbles are removed one at a time without replacement and placed in a row.

a List the possible arrangements.

b How many of these arrangements have a red marble?

Steps	Working
a 1 Draw a tree diagram with 2 sets of branches representing the 2 selections. Once a blue marble has been removed, it cannot appear again.	
2 Each path in the tree from left to right represents a possible arrangement.	
b 3 The arrangements with a red marble are *RR*, *RB*, *RG*, *BR* and *GR*. Count the arrangements with a red marble.	

SB

p. 237

MATCHED EXAMPLE 5 Finding the number of permutations

Five family photos are displayed on a wall. In how many ways can this be done if there are 7 photos to choose from?

Steps	Working
1 There are 5 places to fill, so draw 5 boxes.	
2 There are 7 ways to fill the first position.	
3 Once the first place is filled, there are 6 photos left to fill the second position.	
4 Once the second position is filled, there are 5 photos left to put in the third position.	
5 Once the third position is filled, there are 4 photos left to put in the fourth position.	
6 Once the fourth position is filled, there are 3 photos left to put in the fifth position.	
7 Use the multiplication principle to calculate the number of possible arrangements.	

SB

Using CAS 1:
Permutations
p. 237

MATCHED EXAMPLE 6 | Applying factorials

Ten students are in a line in hostel for dinner. In how many ways can they be served dinner?

Steps	**Working**
We are required to arrange 10 people in a line.	
This can be done in 10! ways.	

SB

Using CAS 2:
Factorials
p. 238

SB

p. 239

MATCHED EXAMPLE 7   Calculating permutations

Use the permutation formula to find the value of

a $^8P_3$ **b** $^7P_4$

Steps	Working
a Substitute $n = 8$ and $r = 3$ into the formula $$^nP_r = \frac{n!}{(n-r)!}$$	
b Substitute into the formula and calculate as the product of 4 consecutive descending numbers starting with 7.	

MATCHED EXAMPLE 8 | Applying permutations with a restriction

The word TRAIN must be arranged in such a way that the vowels occupy the extreme ends. How many such arrangements are possible?

Steps	Working
1 There are 5 letters to arrange, so draw 5 boxes.	
2 The restriction requires the vowels to be arranged at either end. The first and last positions are filled first. There are 2 vowels A and I. The first position can be filled in 2 ways—either the A or the I can be placed in this position. This leaves 1 way to fill the last position.	
3 The other 3 letters can now be placed. There are 3 ways to fill the second position. Once that is filled, there are 2 ways to fill the third position. Continue in the same way until all the positions are filled.	
4 Use the multiplication principle to calculate the number of arrangements.	

MATCHED EXAMPLE 9	Counting arrangements with identical numbers

How many five-digit numbers can be made using three 4s and two 3s?

Steps	**Working**
There are 5 digits to arrange of which 3 digits are of one type (4s) and 2 digits are of a second type (3s). Use the formula $\dfrac{n!}{a! \times b!}$ where $n = 5$, $a = 3$ and $b = 2$.	

MATCHED EXAMPLE 10	Counting arrangements with identical letters

How many arrangements of the letters in the word MISSISSIPPI are possible?

Steps	Working
There are 11 letters to arrange of which there are 4 letter Ss, 2 letter Ps and 4 letter Is. Use the formula $\dfrac{n!}{a! \times b! \times c!}$ where $n = 11$, $a = 4$, $b = 2$ and $c = 4$.	

6

MATCHED EXAMPLE 11 | Applying permutations with groups

How many code words (words with or without meaning) can be formed using all the letters of the word ADMIT at a time so that the vowels occur together?

Steps	Working
1 There are 2 vowels and 3 consonants.	
Put the 2 vowels in a group and count as one object.	
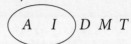	
There are 4 objects to arrange.	
2 Determine the number of ways of arranging the vowels in the group.	
3 Determine the total number of possible arrangements by applying the multiplication principle.	

MATCHED EXAMPLE 12 | Permutations with 2 restrictions

SB

p. 241

A five-digit number is made from the set of digits {1, 2, 3, 4, 7, 9}. How many even numbers greater than 5000 are possible if no repetitions are allowed?

Steps	Working
1 Draw 5 boxes to represent the five-digit number.	
2 Fill the boxes for the restriction first.	
The first box can be filled in 2 ways with the digits 7 or 9. This will produce a number greater than 5000.	
The last box can be filled in 2 ways with the digits 2 or 4. This will produce an even number.	
3 Once these positions are filled, there are 4 numbers remaining. Fill the second box with 4 possible digits, the third box with 3 possible digits and the fourth with 2 possible digits.	
4 Use the multiplication principle to determine the number of possible digits.	

6

SB

p. 244

MATCHED EXAMPLE 13 Calculating combinations

Evaluate $^9C_4$.

Steps	Working
1 In the combination, the numerator is the product of 4 descending numbers starting with 9. The denominator is 4!	
2 Cancel a factor of 8 and evaluate.	

MATCHED EXAMPLE 14	Calculating combinations by symmetry

Evaluate $^{12}C_8$.

Steps	Working
By symmetry $^{12}C_8 = {}^{12}C_4$.	
Evaluate $^{12}C_4$.	

p. 245

6

Using CAS 3:
Combinations
p. 245

MATCHED EXAMPLE 15 | Applying combinations

How many different hands of 7 cards can be dealt from a pack of 52 playing cards?

Steps	Working
The number of different hands of 7 cards is a combination because the order the cards are dealt with does not change the hand. We are choosing 7 cards from 52 cards, which can be done in $^{52}C_7$ ways.	

p. 246

A mixed team of 7 students must be selected from a group of 12 girls and 8 boys for an interschool competition. How many different teams are possible if the team contains 5 girls and 2 boys?

Steps	**Working**
1 The number of different teams is a combination because the order of selection is not important. We are choosing 5 girls from 12 girls AND 2 boys from 8 boys.	
2 Use the multiplication principle to find the total number of teams.	

MATCHED EXAMPLE 17 | Using Pascal's identity

Use Pascal's identity to find possible values of x and y if $^{11}C_6 = {}^{10}C_x + {}^{10}C_y$.

Steps	Working
1 Find the values of n and k in Pascal's identity $$^nC_k = {}^{n-1}C_{k-1} + {}^{n-1}C_k$$	
2 Rewrite Pascal's identity for $n = 11$ and $k = 6$.	
3 Write other equivalent expressions for the combinations found using $^nC_r = {}^nC_{n-r}$.	
4 Write the two possible solutions.	

William can select up to eight different flavoured ice creams. How many choices does he have if he must select at least one?

Steps	Working
Use the rule $^nC_1 + {}^nC_2 + ... + {}^nC_{n-1} + {}^nC_n = 2^n - 1$.	

MATCHED EXAMPLE 19   Applying the pigeon-hole principle

A bag contains four red beads and six blue beads. What is the smallest number of beads that must be drawn from the bag without seeing such that two of them are the same colour?

Steps	Working
The beads are sorted into colours after selection. There are two colours that correspond to two pigeon holes.	
The pigeon-hole principle states that there must be $n + 1$ selections for n categories.	

MATCHED EXAMPLE 20 | Applying the generalised pigeon-hole principle 1

Seventy books of two sizes are sorted into racks as large or small. Find the value of x if at least one rack has at least x books.

Steps	Working
There are 70 books and 2 baskets. At least one rack has at least $\dfrac{70}{2}$ books.	

p. 250

MATCHED EXAMPLE 21 | Applying the generalised pigeon-hole principle 2

Seventeen natural numbers are divided by 4 and the remainder is recorded. Find the value of x if at least x of the numbers have the same remainder.

Steps	Working
1 Determine the possible remainders when a natural number is divided by 4. Each remainder represents a pigeon hole.	
2 Divide the number of objects (17) by the number of holes (4).	

9780170464109

SB

p. 250

Prove that if 8 different numbers are selected from the set $\{1, 2, 3, 4, 5, 6, 7, 8, 9, 10, 11\}$, then 2 of these numbers will have a sum of 12.

Steps	Working
1 Determine the categories for the selected numbers.	
2 The pigeon-hole principle states the following: If n objects are placed in k holes, where $n > k$, then there is at least one hole containing at least $\dfrac{n}{k}$ objects. Divide the number of selected objects (8) by the number of holes (6).	

SB

p. 253

MATCHED EXAMPLE 23 | Applying the inclusion–exclusion principle for 2 sets

Out of a group of 46 students, 26 attend guitar classes and 32 attend piano classes. If every student attends either a guitar class or a piano class, find the number of students who attend both classes.

Steps	Working
1 Let G represent the set of students who attend guitar classes and P represent the set of students who attend piano classes. Write the given information in set notation.	
2 Substitute into the inclusion–exclusion formula.	
3 Write the answer.	

MATCHED EXAMPLE 24 | Finding multiples of 2 numbers in an interval

Find how many integers between 1 and 280 are multiples of 2 or 7.

Steps	Working
1 Let A be the set of numbers that are multiples of 2 and B be the set of numbers that are multiples of 7.	
$A \cap B$ represents integers that are multiple of 2 and 7, that is, multiples of 14.	
Find the number of elements in each of these sets.	
2 Substitute into the inclusion–exclusion formula.	
3 Write the answer.	

MATCHED EXAMPLE 25	Applying the inclusion–exclusion principle for 3 sets

In a class, every student studies mathematics, physics or computer science. The list of entries showed that

- 23 majored in mathematics
- 15 majored in physics
- 28 majored in computer science
- 6 majored in mathematics and physics
- 7 majored in physics and computer science
- 5 majored in mathematics and computer science
- 3 majored in all 3 subjects

Find the number of students who majored in mathematics, physics or computer science.

Steps	Working
1 Let M be the set of students who majored in mathematics, P be the set of students who majored in physics and C be the set of students who majored in computer science. Write the information given in set notation.	
2 Substitute into the inclusion–exclusion formula and answer the question.	

| **MATCHED EXAMPLE 26** | Finding multiples of 3 numbers in an interval |

Find how many integers between 1 and 600 are multiples of 3, 4 or 5.

Steps	**Working**
1 Let A be the set of numbers that are multiples of 3, B be the set of numbers that are multiples of 4 and C be the set of numbers that are multiples of 5.	
$A \cap B$ are multiple of 3 and 4, that is, multiples of 12.	
$A \cap C$ are multiple of 3 and 5, that is, multiples of 15.	
$B \cap C$ are multiple of 4 and 5, that is, multiples of 20.	
$A \cap B \cap C$ are multiple of 3, 4 and 5, that is, multiples of 60.	
Find the number of elements in each set.	
2 Substitute into the inclusion–exclusion formula and answer the question.	

CHAPTER

7 TRIGONOMETRIC IDENTITIES

SB

p. 267

MATCHED EXAMPLE 1 | Arc length, sector and segment

For the circle with 2 radii of length 12 cm forming an angle $\dfrac{\pi}{4}$ radians at the centre of the circle, find:

a arc length AB

b area of the sector AOB

c area of the segment

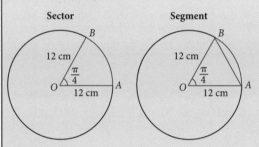

Steps	Working
a Use the formula $l = r\theta$.	
b Use the formula $A = \dfrac{1}{2}r^2\theta$.	
c Use the formula $A = \dfrac{1}{2}r^2(\theta - \sin(\theta))$.	

An exact area is preferred but a decimal approximation helps to compare values.

MATCHED EXAMPLE 2 | The sine rule 1

For this triangle, find the unknown side a correct to two decimal places.

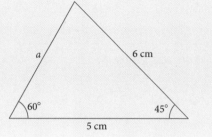

Steps | **Working**

1 Use the sine rule

$$\frac{a}{\sin A} = \frac{b}{\sin B} = \frac{c}{\sin C}$$

2 Solve for a.

MATCHED EXAMPLE 3 | The sine rule for angles 1

For this triangle, find the possible angles B correct to 2 decimal places.

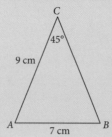

Steps	Working
1 Look for 2 pairs of angles and their opposite sides. Put the unknown variable in the numerator. $$\frac{\sin(C)}{c} = \frac{\sin(B)}{b}$$	
2 Solve for B using solve or $\sin^{-1}$. ┌─────────────────────────────────────┐ There are 2 solutions to the sine equation that are supplementary (add up to 180°). This is because $\sin(x)$ is positive in the 1st and 2nd quadrants. └─────────────────────────────────────┘ **TI-Nspire** **ClassPad**	
3 Test whether the obtuse value can be an angle in a triangle, using the known angle 45°.	

MATCHED EXAMPLE 4 | The sine rule 2

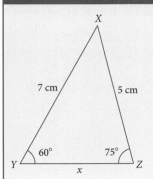

Find length x, correct to one decimal place.

Steps	Working
1 Find $\angle X$.	
2 Use a known side and corresponding angle, with the unknown on top.	
3 Substitute values.	
4 Solve for x.	
5 Evaluate and round appropriately.	

MATCHED EXAMPLE 5 The sine rule for angles 2

Find angle θ, correct to the nearest degree.

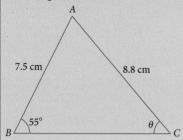

Steps	Working
1 Use a known side and corresponding angle, with the unknown on top.	
2 Substitute values.	
3 Solve for $\sin(\theta)$.	
4 Evaluate.	
5 Use $\sin^{-1}$ to find the acute value of θ.	
6 Find the possible obtuse angle.	
7 Is the obtuse angle possible?	
8 Sum $> 180°$ means A cannot exist for obtuse θ.	

9780170464109

MATCHED EXAMPLE 6 | The sine rule for angles 3

Find Q correct to the nearest 0.1°.

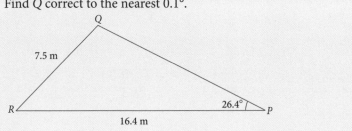

Steps	**Working**
1 Use a known side and corresponding angle, with the unknown on top.	
2 Substitute values.	
3 Solve for $\sin(Q)$.	
4 Evaluate.	
5 Use $\sin^{-1}$ to find the acute value of Q and round.	
6 Find the possible obtuse angle.	
7 Is the obtuse angle possible?	
8 Sum $< 180°$ means C does exist for obtuse Q	

SB

Using CAS 1:
The sine rule
p. 272

MATCHED EXAMPLE 7 | The cosine rule

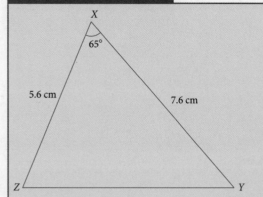

Find the length of YZ correct to two decimal places in the triangle shown.

Steps	Working
1 Use the cosine rule of the form $$x^2 = y^2 + z^2 - 2yz\cos X$$	
2 $\{x = \}$.	
3 Select +ve x.	
4 Write the answer.	

MATCHED EXAMPLE 8 | The cosine rule for angles

Find the angle C in the triangle shown, to the nearest degree.

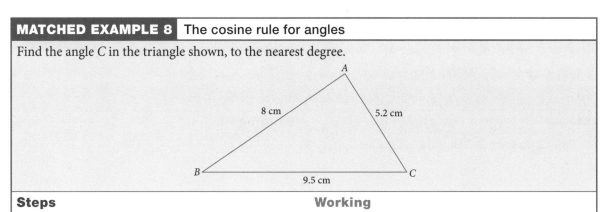

Steps	Working
1 Use the cosine rule.	
2 Solve for angle C, restricting C to $0 < C < 180°$.	
3 Write the answer to the nearest degree.	

SB

Using CAS 2:
The cosine rule
p. 276

SB

p. 279

MATCHED EXAMPLE 9 | Angles of elevation

Jermaine sees that the top of a sand dune at the end of a beach is at an angle of elevation of 22°.

On walking 50 m closer, she finds that the angle of elevation increases to 32°.

What is the approximate height of the sand dune correct to one decimal place?

Steps	Working
1 Draw a diagram and label the sides and angles.	
2 Use the exterior angle of △*SBT* to find ∠*STF*.	
3 Use the sine rule.	
4 Solve for *x*.	
5 Now use △*SBT* to find *h* from *x*.	
6 Solve for *h*.	
7 Substitute the expression for *x* and calculate the answer.	
8 Round appropriately and write the answer.	

MATCHED EXAMPLE 10 | Bearings

Two fire towers in the direction N 52° W and at the bearing of 98° are at ranges of 5.67 km and 9.88 km respectively. Find the distance between the towers.

Steps	Working
1 Draw and label a diagram showing the information. The angle between the directions of the towers is 46°.	
2 Write the cosine rule for the desired side.	
3 Substitute values.	
4 Calculate and find p.	
5 Round appropriately and write the answer.	

SB

p. 280

MATCHED EXAMPLE 11 Bearings with angles of elevation 1

Walter observes that the top of a building at a bearing of 035° is at an angle of elevation of 10°. Samuel is 400 m due east of Walter, and he says the bearing of the building is 290°.

Find correct to the nearest metre the height of the building.

Steps	Working
1 Draw a diagram in a sketch of a box to make it look 3D and label the sides and angles.	
2 Find ∠SWB and ∠BSW.	
3 'Unfold' and draw the triangles flat.	
4 Find ∠WBS.	
5 Use the sine rule to find x.	
6 Solve for x.	
7 Now use △WTB to find h from x.	
8 Solve for h.	
9 Substitute the expressions for x.	
10 Round appropriately and write the answer.	

MATCHED EXAMPLE 12 | Bearings with angles of elevation 2

From a certain point A, the top of a tower due north has an angle of elevation of 18°. From another point, B, 3 km east of A and on the same level as A, the bearing of the tower is N 42° W. Find the height of the tower above the level of A and B.

Steps	Working
1 Draw a diagram in a sketch of a box to make it look 3D. Call the foot of the tower W and the top T. Show directions of north and east.	
2 Investigate known angles.	
3 Call the height of the tower h and the distance from A to the foot of the tower x. Redraw the diagram with the sides and angles marked, ignoring the parts of the box you don't need.	
4 'Unfold' and draw the triangles flat. In this case, they are joined along AW.	

5 Use tan in $\triangle WAB$ to find x and rearrange.

6 Use tan in $\triangle TAW$ to find h.

7 Substitute x.

8 Calculate the answer.

9 Round appropriately and write the answer.

MATCHED EXAMPLE 13 | Exact values 1

Find the exact value of each trigonometric ratio.

a $\cot(60°)$ **b** $\sec(30°)$ **c** $\text{cosec}(45°)$

Steps	Working
a Write the reciprocal function cot, in terms of the original function tan, evaluate and simplify.	
b Write the reciprocal function sec, in terms of the original function cos, evaluate and simplify.	
c Write the reciprocal function cosec, in terms of the original function sin, evaluate and simplify.	

MATCHED EXAMPLE 14 | Exact values 2

Evaluate $\sec\left(\dfrac{4\pi}{3}\right)$.

Steps	Working
1 sec is the reciprocal of cos.	
2 $\dfrac{4\pi}{3}$ is in the 3rd quadrant, where cos is negative. Its reference angle is $\dfrac{\pi}{3}$.	

9780170464109

MATCHED EXAMPLE 15 Pythagorean identities

Use a trigonometric identity to find $\operatorname{cosec}(x)$ for $\dfrac{\pi}{2} \leq x \leq \pi$ if $\cot(x) = -\dfrac{5}{3}$.

Steps	Working
1 Use the trigonometric identity $1 + \cot^2(\theta) = \operatorname{cosec}^2(\theta)$ and solve for $\operatorname{cosec}^2(\theta)$.	
2 Given $\dfrac{\pi}{2} \leq x \leq \pi$ in the 2nd quadrant, sin is positive and so is cosec.	

Alternative Method

1 Draw a right-angled triangle using $\cot(x) = \dfrac{5}{3}$, using Pythagoras' theorem to find the hypotenuse. Don't consider the negative at this stage, as this is irrelevant for the sides of a triangle.	
2 Find $\operatorname{cosec}(x)$.	
3 But consider the quadrant. Given $\dfrac{\pi}{2} \leq x \leq \pi$ in the 2nd quadrant, sin is positive and so is cosec.	

7

MATCHED EXAMPLE 16 — Compound angle formulas 1

Expand $\tan (x + 3y)$.

Steps	Working
Use the trigonometric identity $$\tan (x + y) = \frac{\tan(x) + \tan(y)}{1 - \tan(x)\tan(y)}$$	

SB
p. 291

Evaluate $\sin\left(\dfrac{5\pi}{12}\right)$.

Steps	Working
1 $\dfrac{5\pi}{12} = \dfrac{2\pi}{12} + \dfrac{3\pi}{12} = \dfrac{\pi}{6} + \dfrac{\pi}{4}$	
2 Use the trigonometric identity $\sin(x+y) = \sin(x)\cos(y) + \cos(x)\sin(y)$.	
3 Evaluate using exact values.	

SB

Using CAS 3:
The sum and
difference
identities
p. 291

MATCHED EXAMPLE 18 | Compound angle formulas 3

Use a suitable compound angle formula to find the expansion of $\sin\left(\dfrac{\pi}{2} + x\right)$.

Steps	Working
1 Use the trigonometric identity $\sin(x + y) = \sin(x)\cos(y) + \cos(x)\sin(y)$.	
2 Evaluate and simplify.	

MATCHED EXAMPLE 19	Double angle formula 1

Use a double angle formula to find an expression for $\cos(8x)$ in terms of $4x$.

Steps	Working
1 Express $8x$ as $4x + 4x$.	
2 Use the identity $\cos(2x) = 2\cos^2 x - 1$ to find an expression for $\cos(8x)$.	

MATCHED EXAMPLE 20 | Double angle formula 2

Use a suitable double angle formula to find an expression for $\sin\left(\dfrac{3\pi}{4}\right)$.

Steps	Working
1 Find an expression for a double angle.	
2 Use the identity $\cos(2x) = 1 - 2\sin^2(x)$ using $\dfrac{3\pi}{2} = 2 \times \dfrac{3\pi}{4}$	
3 Rearrange to get an expression for $\sin\left(\dfrac{3\pi}{4}\right)$.	

9780170464109

SB

p. 298

Prove that $\cos(x)\cos(y) = \dfrac{1}{2}\left[\cos(x-y)+\cos(x+y)\right]$.

Steps	Working
1 Expand and simplify the RHS using the identities for $\cos(x+y)$ and $\cos(x-y)$.	
2 Show that it is equal to the LHS.	

7

MATCHED EXAMPLE 22 | Product of cosines 2

Express $\cos(3x)\cos(5x)$ as a sum or a difference of trigonometric functions.

Steps	Working
1 Use the formula $$\cos(x)\cos(y) = \frac{1}{2}\left[\cos(x-y) + \cos(x+y)\right]$$	
2 Collect terms and simplify.	
3 Remember that $\cos(-\theta) = \cos(\theta)$.	
4 Write the result.	

MATCHED EXAMPLE 23 | Graphing $y = a \cos(x) + b \sin(x)$

Express $2\sin(x) + 3\cos(x)$ in the form $r\sin(x + \alpha)$ and sketch the graph of $y = 2\sin(x) + 3\cos(x)$ for $x = 0$ to 2π.

Steps	Working
1 Expand the RHS using the sum identity.	
2 Let $2\sin(x) + 3\cos(x) = r\sin(x + \alpha)$.	
3 Compare the LHS and RHS. These are simultaneous equations in r and α.	
4 To find r, square both equations and add them.	
5 To find α, divide $r\sin(\alpha)$ by $r\cos(\alpha)$.	
6 Write the result.	
7 Sketch a sine graph with amplitude $\sqrt{13}$, period 2π and a y-intercept at $2\sin(0) + 3\cos(0) = 3$.	

SB

p. 301

MATCHED EXAMPLE 24 Solving $a \cos(x) + b \sin(x) = c$

Express $\cos(x) - \sin(x)$ in the form $r \cos(x + \alpha)$ and hence solve the equation $\cos(x) - \sin(x) = \sqrt{2}$
for $x = 0$ to 2π.

Steps	Working
1 Find $r = \sqrt{a^2 + b^2}$.	
2 Use $\tan(\alpha) = \left\lvert \dfrac{b}{a} \right\rvert$ to find α.	
3 Write $\cos(x) - \sin(x)$ in the form $r \cos(x + \alpha)$.	
4 Solve the equation $\cos(x) - \sin(x) = \sqrt{2}$.	

9780170464109

MATCHED EXAMPLE 25 | Proving trigonometric identities 1

Prove the identity $2\tan^2(x)\cos^2(x) = 1 - \cos(2x)$.

Steps	Working
Start with the RHS and use the double angle formula $\cos(2x) = 1 - 2\sin^2 x$.	
Simplify the LHS using the identity $\tan(x) = \dfrac{\sin(x)}{\cos(x)}$.	
Compare LHS and RHS.	

SB

Using CAS 4:
Verifying
trigonometric
identities
p. 303

MATCHED EXAMPLE 26 | Proving trigonometric identities 2

Prove the identity $\dfrac{\sin(a)}{1+\cos(a)} = \dfrac{\frac{1}{2}[1-\cos(2a)}{\sin(a)+\frac{1}{2}(\sin 2a)}$.

Steps	Working
1 Start with the RHS by simplifying the fraction.	
2 Take out $\sin(a)$ as a common factor and cancel.	
3 Compare LHS and RHS.	

MATCHED EXAMPLE 27 | Solving trigonometric equations

Use a double angle identity to solve the equation $\cos(2a) = \cos^2(a) - 1$ for $a \in [0, 2\pi]$.

Steps	Working
1 Simplify the LHS using the identity $\cos(2a) = 2\cos^2(a) - 1$.	
2 Take all terms to the LHS and simplify.	
3 Solve the equation for $a \in [0, 2\pi]$.	

7

GRAPHING FUNCTIONS AND RELATIONS

p. 317

MATCHED EXAMPLE 1	Relations, functions, domain and range

State whether each of the following is a function and state the domains, co-domains and ranges.

a $f: R \to R, f(x) = \sqrt{x+2}$

b $g: [0, 6) \to R^+, x + 2 < y \le 3x + 2$

c $(x-2)^2 + (y+5)^2 = 25$

Steps	Working
a 1 Is it a function?	
2 What values can x take?	
3 What are allowable y values?	
4 What set are the actual values?	
b 1 Find the ends of the lines. Sketch the graph, the area above $y = x + 2$ and under or on $y = 3x + 2$ between $x = 0$ and $x = 6$ inclusive of 0. Show inclusion of boundaries by solid lines and exclusion by dashed lines.	
2 Is it a function?	
3 What values can x be?	
4 What are allowable y values?	
5 What are the actual values?	

9780170464109

c 1 State the shape.

   2 Sketch the graph.

   3 Is it a function?

   4 What values can x be?

   5 What are allowable y-values?

   6 What are the actual values?

MATCHED EXAMPLE 2 | Transformations of functions

State the basic function and transformations for each function. State the equation for part c.

a $f(x) = 5\sqrt[3]{4-x} + 1$

b $g(x) = -(2(x+4))^4 - 2$

c

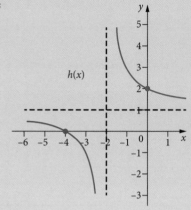

Steps	Working
a 1 State the basic function.	
2 State the transformations by comparing with $af(n(x+b)) + c$.	
b 1 State the basic function.	
2 State the transformations by comparing with $af(n(x+b)) + c$.	
c 1 State the basic function.	
2 State the obvious transformations.	
3 Use the y-intercept $f(0) = 2$ to check for a dilation.	
4 State all the transformations and equation.	

SB
p. 320

Sketch each relation.

a $f(x) = \pm\sqrt{x-5} - 1$

b $g(x) = 4 - \dfrac{1}{(x-6)^2}$

c $h(x) = (2 - 0.25x)^3 + 5$

Steps	Working
a 1 State the basic relation and transformations by comparing with $af(n(x+b)) + c$.	
2 State any important points.	
3 Sketch the graph, marking the important points.	
b 1 State the basic relation and transformations by comparing with $af(n(x+b)) + c$.	
2 State any important points.	
3 Sketch the graph, marking the important points.	

c 1 State the basic relation and transformations by comparing with $af(n(x + b)) + c$.

2 State any important points.

3 Sketch the graph, marking the important points.

9780170464109

MATCHED EXAMPLE 4 | Identities and polynomial factors

$x - 2$ is a factor of $3x^3 - x^2 - 7x - 6$. Use an identity to find the other factor.

Steps	Working
1 Write the identity for the factor $x - 2$.	
2 Expand the brackets.	
3 Equate the coefficients of the polynomial.	
4 Solve for a, b and c.	
Use the 3rd equation to check the values: $c - 2b = 3 - 10 = -7$ ✓	
5 Write the other factor.	

8

MATCHED EXAMPLE 5 | Identities and rational functions

Use an identity to express $\dfrac{5x^2 + 4x - 2}{x - 4}$ in the form $\dfrac{A}{x - 4} + Bx + C.$

Steps	Working
1 Write the desired form.	
2 Change to a common denominator.	
3 Simplify the numerator.	
4 Write the identity.	
5 Equate coefficients of the numerator.	
6 Solve for A, B and C.	
7 Write in the desired form.	

MATCHED EXAMPLE 6 | Rational functions with simple denominators

Express $\dfrac{8x+7}{(x-1)(x+2)}$ in the form $\dfrac{A}{x-1}+\dfrac{B}{x+2}$.

Steps	**Working**
1 Write the desired form.	
2 Change to a common denominator.	
3 Write the numerator identity.	
4 Choose $x=1$ and solve for A.	
5 Choose $x=-2$ and solve for B.	
6 Write the desired form.	
7 If you have time, check the result by expressing with a common denominator.	

MATCHED EXAMPLE 7 | Partial fractions with decreasing powers

Write $\dfrac{2x^2+5x+1}{(x+2)^3}$ as a sum of partial fractions.

Steps	Working
1 Write as a sum of partial fractions with decreasing powers.	
2 Express with a common denominator and simplify.	
3 Equate coefficients of the numerator and solve for A, B, C.	
4 Write as partial fractions.	

MATCHED EXAMPLE 8 | Partial fractions with a quadratic denominator

Write $\dfrac{-10x^2-10x-8}{(x^2+4x+5)(x-1)}$ in the form $\dfrac{Ax+B}{x^2+4x+5}+\dfrac{C}{x-1}$.

Steps	Working
1 Write the desired form.	
2 Change to a common denominator.	
3 Simplify the numerator.	
4 Write the identity.	
5 Equate coefficients of the numerator and solve for A, B and C.	
6 Write the desired form.	

MATCHED EXAMPLE 9 | Partial fractions of rational functions

Write each of the following as an expression with partial fractions, without finding the numerators.

a $\dfrac{8x^3+7x+4}{(x+5)(x+8)^2(x+7)}$
b $\dfrac{x^2+5}{(x+1)(5x^2+4x+2)(4x+7)}$
c $\dfrac{x^2+2x+5}{(x+8)(x+4)}$

Steps	Working
a The denominator has two simple linear factors and a square linear factor.	
Parts a and b have denominators with degree 4 and numerators with a lower degree, so they are simple rational expressions. There is no quotient.	
b 1 Check the quadratic in the denominator.	
2 There are also two simple linear factors in the denominator.	
c 1 Check the degrees of the denominator and numerator.	
2 The denominator has two simple linear factors.	

SB

p. 326

Write $\dfrac{4x^2+4x-4}{(x-1)(x+1)^2}$ as partial fractions.

Steps	Working

1 There are two linear factors in the denominator, one squared.

2 Change to a common denominator.

3 Numerators are identically equal.

4 Choose $x = 1$ and solve for A.

5 Choose $x = -1$ and solve for B.

6 Choose $x = 0$ and substitute A and B to find C.

SB

Using CAS 1:
Partial fractions
p. 327

7 Write the final expression.

MATCHED EXAMPLE 11 | Reciprocal linear functions

Sketch the graph of $f(x) = \dfrac{1}{4-x}$.

Steps	Working
1 Consider where the linear denominator $4-x$ is zero.	
2 Consider the signs of $4-x$.	
3 Find the y-intercept.	
4 Consider the behaviour as $x \to \pm\infty$.	
5 Sketch the graph.	

Sketch the graph of $f(x) = \dfrac{5}{x^2 - 4x + 5}$.

Steps	Working
1 Consider whether the quadratic denominator $x^2 - 4x + 5$ has zeros.	
2 Find the minimum of $x^2 - 4x + 5$ by completing the square or using $x = \dfrac{-b}{2a}$.	
3 Multiply the numerator of $f(x)$ by the y value of the reciprocal of the minimum of the denominator to get the maximum.	
4 Find the y-intercept of $f(x)$.	
5 Consider the behaviour as $x \to \pm\infty$.	
6 Sketch the graph.	

p. 333

MATCHED EXAMPLE 13 · Reciprocal quadratic functions with real zeros

Sketch the graph of $f(x) = \dfrac{5}{4x - x^2 + 21}$.

Steps	Working
1 Consider whether the quadratic denominator $2x - x^2 + 8$ has zeros.	
2 Consider the sign(s).	
3 Find the maximum of $4x - x^2 + 21$ by completing the square or using $x = \dfrac{-b}{2a}$.	
4 Multiply the numerator of $f(x)$ by the y value of the reciprocal of the maximum of the denominator to get the minimum.	
5 Find the y-intercept of $f(x)$.	
6 Consider the behaviour as $x \to \pm\infty$.	
7 Sketch the graph.	

Sketch the graph of $f(x) = \dfrac{36}{(x+3)(x-2)^2(x-5)}$.

SB

p. 334

Steps	Working
1 Consider whether any zeros of the polynomial denominator.	
2 Consider the sign(s).	
3 Consider the behaviour as $x \to \pm\infty$.	
4 Find the y-intercept.	
5 Sketch the graph.	

MATCHED EXAMPLE 15 | Reciprocal polynomial functions 2

The graph of $f(x)$ is shown on the right.

Sketch the graph of $\dfrac{1}{f(x)}$.

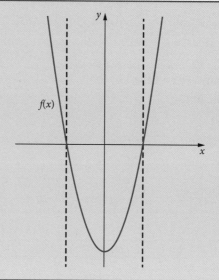

Steps	Working
1 Look for the relevant features of $f(x)$.	
2 Put vertical asymptotes on the new graph instead of the zeros, a zero instead of the vertical asymptote, a minimum instead of a maximum and a maximum instead of a minimum on the graph of $\dfrac{1}{f(x)}$. Show $\dfrac{1}{f(x)} \to 0$ with the same sign where $f(x) \to \pm\infty$.	

3 Join up the lines, making sure the new graph is concave up where the original is concave down and vice versa.

Make sure that the signs are the same.

Write the function name.

MATCHED EXAMPLE 16 | **Exact values of sec, cosec and cot**

Find the values of

a $\sec\left(\dfrac{\pi}{6}\right)$ **b** $\csc\left(\dfrac{5\pi}{4}\right)$ **c** $\cot\left(\dfrac{5\pi}{2}\right)$

Steps	Working
a 1 State the value of $\cos\left(\dfrac{\pi}{6}\right)$. **2** Find the reciprocal.	
b 1 Use CAST for $\sin\left(\dfrac{5\pi}{4}\right)$. **2** Find the value of $\sin\left(\dfrac{5\pi}{4}\right)$. **3** Find the reciprocal.	
c 1 For multiples of $\dfrac{\pi}{2}$, find the coordinates on the unit circle. **2** Use the definition.	

MATCHED EXAMPLE 17 | Exact values using identities

$\operatorname{cosec}(x) = -\sqrt{2}$ and $\pi < x < \dfrac{3\pi}{2}$. Find $\sec(x)$ and $\cot(x)$.

Steps	Working
1 Use a Pythagorean identity to find $\cot(x)$.	
2 Use another identity to find $\sec(x)$.	
3 Use the definition of $\cot(x)$.	
4 Write the answer.	

p. 341

MATCHED EXAMPLE 18 | Sketching a graph as a reciprocal function

Sketch the graph of $f(x) = 2\csc\left(2x + \dfrac{\pi}{6}\right) - 1$ for $-\pi \le x \le \pi$.

Steps	Working
1 At the asymptotes, $\sin\left(2x + \dfrac{\pi}{6}\right) = 0$.	
2 Restrict to $-\pi \le x \le \pi$.	
3 The minima of f are at the maxima of $\sin(x)$, so $\sin\left(2x + \dfrac{\pi}{6}\right) = 1$.	
4 Restrict to $-\pi \le x \le \pi$.	
5 The maxima of f are at the minima of $\sin(x)$: $\sin\left(2x + \dfrac{\pi}{6}\right) = -1$.	
6 Restrict to $-\pi \le x \le \pi$.	

9780170464109

7 Place the important features and complete the sketch.

8

MATCHED EXAMPLE 19 | Sketching a graph using the basic function

Sketch the graph of $y = 3\cot\left(2x - \dfrac{\pi}{2}\right) + 2$ for $0 \le x \le \pi$.

Steps	Working
1 Identify the basic function.	
2 Write as $a\,f(n(x + b)) + c$.	
3 State the transformations.	
4 Find the asymptotes, working from the easiest one first.	
5 Find the centres, working from the easiest zero first.	
6 Find what becomes of the points where the y values are ± 1 on the basic function, working from the easiest one first.	
7 Sketch the graph, including asymptote equations and important coordinates.	

9780170464109

MATCHED EXAMPLE 20 | Identifying the equation of a graph

Part of the graph of $y = f(x)$ is shown.

State a possible equation for the graph.

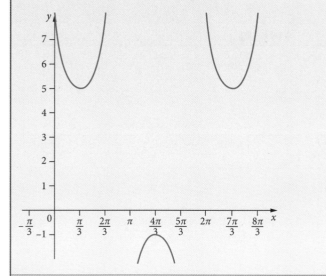

Steps	Working
1 Identify the basic function.	
2 Identify the period and dilation from the y-axis.	
3 Identify the centre line to find any translation parallel to the y-axis.	
4 Use the distance of the minima or maxima.	
5 Use a maximum or minimum to find the phase shift.	
6 Write the equation.	

p. 348

MATCHED EXAMPLE 21 | Values of inverse circular functions

Find the values of

a $\cos^{-1}\left(-\dfrac{\sqrt{3}}{2}\right)$ **b** $\arctan\left(\dfrac{1}{\sqrt{3}}\right)$ **c** $\sin^{-1}\left(-\dfrac{1}{2}\right)$

Steps	Working
a 1 Find $0 \le x \le \pi$ such that $\cos(x) = -\dfrac{\sqrt{3}}{2}$.	
2 Write the answer.	
b 1 Find $-\dfrac{\pi}{2} < x < \dfrac{\pi}{2}$ such that $\tan(x) = \dfrac{1}{\sqrt{3}}$.	
2 Write the answer.	
c 1 Find $-\dfrac{\pi}{2} \le x \le \dfrac{\pi}{2}$ such that $\sin(x) = -\dfrac{1}{2}$.	
2 Write the answer.	

9780170464109

What are the implied domain and range of these functions?

a $f(x) = 2\arccos\left(\dfrac{x+2}{2}\right) + 3$ **b** $g(x) = 2\sin^{-1}(x-1) - 4$

SB
p. 350

Steps	Working
a 1 What do the domain ends −1 and 1 of arccos (x) become?	
2 What is the range?	
3 Write the answer.	
b 1 What is the implied domain?	
2 What happens to the y-values of $\sin^{-1}(x)$?	
3 Write the answer.	

MATCHED EXAMPLE 23	Sketching an arccosine function

Sketch the graph of $f(x) = 2\cos^{-1}\left(2 - \dfrac{x}{3}\right) + 1$.

Steps	Working
Method 1: Find important points using transformations	
1 Write as $af(n(x + b)) + c$.	
2 State the transformations.	
3 Find the ends and centre.	
Method 2: Find important points by analysis	
1 Find where $\cos^{-1}(x) = 1, 0$ and -1.	
2 Substitute to find points.	
Both methods	
3 Put the important points on the graph and join to sketch the graph.	

MATCHED EXAMPLE 24 | Sketching an arctangent function

Sketch the graph of $f(x) = \tan^{-1}(4x + 3) + 3$.

Steps	Working
1 Write as $af(n(x + b)) + c$.	
2 State the transformations.	
3 Find the important points.	
4 State the asymptotes.	
5 Find the y-intercept.	
6 Put in the important points and equations and sketch the graph.	

8

MATCHED EXAMPLE 25 | Absolute values

Simplify each of the following, writing without | | signs.

a |−25| **b** |54| **c** |58 − 74|

Steps	Working
a Use the absolute value function.	
b Use the absolute value function.	
c Simplify and use the absolution value function.	

MATCHED EXAMPLE 26	Equations with absolute values

Solve $|11 - 2m| = 8m + 40$.

Steps	Working
1 Split into two equations.	
2 Solve each equation.	

MATCHED EXAMPLE 27 | Inequalities with absolute values

Solve

a $|8x - 3| < 13$ **b** $|5 - x| \geq 27$ **c** $|7x - 4| > 2x + 1$

d $|4 - 8x| \leq 8x - 10$ **e** $|4x - 1| > 8x + 10$

Steps	Working
a **1** Split into two inequalities.	
2 Solve each inequality.	
3 Write the solution.	
b **1** Split into two inequalities.	
2 Solve each inequality.	
c **1** Split into two inequalities.	
2 Solve each inequality.	
3 Write the solution.	
d **1** Split into two inequalities.	
2 Solve each inequality.	
3 Write the solution.	
e **1** Split into two inequalities.	
2 Solve each inequality.	
3 Write the solution.	

9780170464109

Sketch each of the following for the stated domain.

a $f(x) = -5|x| + 2, -5 < x < 5$

b $y = |x - 2| + 1, -2 < x < 7$

c $f(x) = \dfrac{|x + 2| - 3}{4}, -6 < x < 4$

Steps	Working
a 1 State the transformations.	
2 State the positions of some points.	
3 Sketch the graph.	
b 1 State the transformations.	
2 State the positions some points.	
3 Sketch the graph.	
c 1 Write as $af(n(x + b)) + c$.	
2 State the transformations.	
3 State the positions some points.	

SB

p. 356

8

4 Sketch the graph.

MATCHED EXAMPLE 29 | A simple locus

Find the equation of the locus of points equidistant from $A(4, 2)$ and $B(6, -6)$ and describe the locus.

Steps	Working
1 Sketch A, B and some possible points $P(x, y)$ on the locus.	
2 Write the condition.	
3 Use the distance formula.	
4 Simplify.	
5 It looks perpendicular to AB gradient.	
6 Find the product $m_1 m_2$.	
7 Find the midpoint of AB and check if it is on locus.	
8 Write the answer.	

MATCHED EXAMPLE 30 | A locus involving angles

A and B have coordinates $(-2, 3)$ and $(4, 4)$, respectively. Find the equation of the locus of point P such that $\angle APB = 90°$ and describe the locus.

Steps	Working
1 Sketch A, B and some possible points $P(x, y)$ on the locus. The points look like they could make a circle.	
2 Write the condition.	
3 Use gradients with $P(x, y)$.	
4 Simplify.	
5 Complete squares.	
6 Consider the equation.	
7 Consider the endpoints.	
8 Write the answer.	

9780170464109

MATCHED EXAMPLE 31	Centre and radius of a circle

Find the centre and radius of the circle with equation $4x^2 - 16x + 4y^2 + 24y = 36$.

Steps	**Working**
1 Make the coefficients of x^2 and y^2 both 1.	
2 Complete the squares.	
3 Write as squares.	
4 Write in standard form.	
5 Write the answer.	

MATCHED EXAMPLE 32 | Horizontal parabola

Sketch the graph of $(y + 2)^2 = -4(x - 3)$.

Steps	Working
1 Write the vertex.	
2 Write the other characteristics.	
3 Find any intercepts.	
4 Sketch the graph.	

MATCHED EXAMPLE 33 | Vertical parabola

Sketch the graph of $x^2 - 2y - 8x + 10 = 0$.

Steps	Working
1 Express in standard form by completing the square.	
2 Write the vertex.	
3 Write the other characteristics.	
4 Find any intercepts.	
5 Sketch the graph.	

8

MATCHED EXAMPLE 34 | Horizontal ellipse

Sketch the graph of $4x^2 + 16y^2 = 16$.

Steps	Working
1 Write in standard form.	
2 Write a and b.	
3 Find e.	
4 Find the foci.	
5 Sketch the graph, showing the intercepts and foci.	

MATCHED EXAMPLE 35 | Horizontal hyperbola

Sketch the graph of $\dfrac{x^2}{9} - \dfrac{y^2}{16} = 1$.

Steps	Working
1 Find the values of a and b.	
2 Find e and the foci.	
3 Sketch the auxiliary rectangle and use the corners to sketch the asymptotes.	
4 Use the asymptotes and auxiliary rectangle to sketch the hyperbola and mark the vertices and foci.	

MATCHED EXAMPLE 36 | Vertical hyperbola

Find the equation and sketch the graph of a hyperbola with vertices at (3, 2) and (3, −4) and an asymptote given by $y = 2x - 7$.

Steps	Working
1 Make a rough sketch, showing the vertices and the asymptote. The vertices are in a vertical line, so the transverse axis must be vertical.	
2 Find the centre, a, b, e and the foci.	
3 Substitute a, b, h, k for the equation and simplify.	
4 Sketch the graph.	

MATCHED EXAMPLE 37 | Polar to Cartesian form of an ellipse

Find the equation of $r = \dfrac{12}{4 + 2\sin(\theta)}$ in Cartesian form and sketch the graph.

Steps	Working
1 Write the polar equation in standard form to find e.	
2 State the kind of conic.	
3 Substitute $\theta = \dfrac{\pi}{2}$ and $\dfrac{3\pi}{2}$ to find the major vertices.	
4 Change to Cartesian form and find (h, k) and a.	
5 Find the value of b.	
6 Write the standard form of the vertical Cartesian equation.	
7 Substitute values, taking account of the translation of the centre.	
8 Simplify to write the answer.	
9 Sketch the graph.	

SB

p. 374

MATCHED EXAMPLE 38 | Cartesian to polar form of an ellipse

Express $\dfrac{(x-3)^2}{25} + \dfrac{y^2}{16} = 1$ in polar form.

Steps	Working
1 State the type of conic.	
2 Find the value of e.	
3 Check that a focus is at $(0, 0)$.	
4 Find the value of d.	
5 Decide on the equation form.	
6 Write the general polar equation.	
7 Substitute the values and simplify.	
8 Write the answer.	

MATCHED EXAMPLE 39	Cartesian to parametric form of an ellipse

Write the equation of $\dfrac{(x-4)^2}{36}+\dfrac{(y+1)^2}{9}=1$ in parametric form.

Steps	Working
1 State the type of conic.	
2 Write the values of a and b.	
3 Write the parametric form, including an adjustment for the translation.	

8

MATCHED EXAMPLE 40 | Polar to Cartesian form of a hyperbola

Write the equation of $r = \dfrac{3}{1 + 2\cos(\theta)}$ in Cartesian form and sketch the graph.

Steps	Working
1 Write in standard form to find the value of e.	
2 Use the value e to identify the conic.	
3 Find the vertices.	
4 Write the vertices in Cartesian form.	
5 Find the centre (the midpoint of VV') and the value of a ($VV' = 2a$).	
6 Find the value of b.	
7 Write the standard equation of the horizontal hyperbola.	
8 Substitute values.	
9 Simplify.	
10 Sketch the graph.	

MATCHED EXAMPLE 41 Cartesian to polar form of a hyperbola

Write the equation of $\dfrac{(2y-8)^2}{36} - \dfrac{x^2}{7} = 1$ in polar form.

Steps	Working
1 Write in standard form to find the values of a and b.	
2 Write the values of a and b.	
3 Find the value of e.	
4 State the centre.	
5 Check that a focus is at $(0, 0)$.	
6 Find the value of d.	
7 Write the standard polar form for a vertical hyperbola with the lower focus at $(0, 0)$.	
8 Write the answer.	

MATCHED EXAMPLE 42 | Parametric to Cartesian form of a hyperbola

Write the pair of equations $\begin{cases} x = \dfrac{3}{\cos(\theta)} \\ y = 5\tan(\theta) \end{cases}$ in cartesian form.

Steps	Working
1 State the type of conic.	
2 Write the values of a and b.	
3 Write the standard cartesian form.	
4 Simplify and write the answer.	

Write the equation of $r = \dfrac{3}{2 - \cos(\theta)}$ in Cartesian form and sketch the graph.

SB
p. 378

Steps	Working
1 State the type of conic.	
2 Find the vertex and other characteristics.	
3 Find the value of a.	
4 Write the equation.	
5 Substitute values.	
6 Find the intercepts.	
7 Sketch the graph.	

8

MATCHED EXAMPLE 44 | Parametric to Cartesian form of a parabola

Write the equation of $\begin{cases} x = 2t^2 - 1 \\ y = 4t + 3 \end{cases}$ for $t \in R$ in Cartesian form and sketch the graph.

Steps	Working
1 State the type of function.	
2 What other characteristics has it?	
3 Write the equation in cartesian form.	
4 Find the y-intercepts.	
5 Sketch the graph.	

MATCHED EXAMPLE 45 | Parametric to Cartesian form of a circle

Write the equation of $\begin{cases} x = 4\cos(\theta) - 1 \\ y = 4\sin(\theta) + 2 \end{cases}$ for $0 \leq \theta < 2\pi$ in cartesian form and sketch the graph.

Steps	Working
1 State the type of function.	
2 Write the centre.	
3 Write the equation.	
4 Sketch the graph.	

8

CHAPTER

9 COMPLEX NUMBERS

SB

p. 394

MATCHED EXAMPLE 1	Finding the square root of a negative number

Use the fact that $i = \sqrt{-1}$ to simplify each root in terms of i.

a $\sqrt{-16}$ **b** $\sqrt{-25}$ **c** $\sqrt{-72}$

Steps	Working
a Write -16 as a product of 16 and -1.	
b Write -25 as a product.	
c Write -72 as a product, then simplify.	

SB

Using CAS 1:
Imaginary
numbers
p. 394

MATCHED EXAMPLE 2 | Simplify expressions involving powers of i

Use $i^2 = -1$ to simplify each expression.

a $2i^2 + 3i^5$ **b** $\dfrac{2i^2 - 8i^7}{2i^2}$ **c** $\dfrac{\sqrt{i^{4n}} \times \sqrt{i}}{i^{2n}}$

Steps	Working
a Factorise i^2 in the second term to turn into -1.	
b Factorise the numerator and simplify.	
c 1 Rewrite in index form.	
2 Simplify using index laws.	

MATCHED EXAMPLE 3	The real part and the imaginary part of a complex number

Find Re (z) and Im (z) for each complex number.

a $z = 9 + 6i$ **b** $z = \sqrt{5}i - 8$

c $z = 3x - 7y + (5x - 8y)i$ **d** $z = \dfrac{7x + 4yi}{x^3 - y^3}$

Steps	Working
a Identify the real and imaginary parts.	
b $z = \sqrt{5}i - 8$ is the same as $z = -8 + \sqrt{5}i$.	
c Identify the real and imaginary parts.	
d Write $z = \dfrac{7x + 4yi}{x^3 - y^3}$ as $z = \dfrac{7x}{x^3 - y^3} + \left(\dfrac{4y}{x^3 - y^3}\right)i$.	

Using CAS 2:
Real and imaginary
parts of a complex
number
p. 396

SB

p. 397

MATCHED EXAMPLE 4 | Adding and subtracting complex numbers

Simplify each expression and write the result in the form $z = a + bi$.

a $-7i + 8 + 2i - 4$

b $\dfrac{3}{4}(16 - 24i) - \dfrac{i}{2}(4 - 4i)$

c $-8i - \dfrac{4}{3i}$

Steps	Working
a 1 Group the real and imaginary parts.	
2 Simplify.	
b Expand and simplify.	
c Simplify the fraction by multiplying numerator and denominator by **i**. This is called **realising the denominator**: making the denominator a real number.	

MATCHED EXAMPLE 5 | Multiplying complex numbers

Simplify each expression.

a $(-4-i)(5+4i)$ **b** $13-(1+3i)^2$ **c** $-7i(3-4i)(3+4i)$

Steps	Working
a **1** Expand the binomial product.	
2 Simplify using $i^2 = -1$.	
b Expand the perfect square and simplify.	
c Expand and simplify, using difference of 2 squares.	

MATCHED EXAMPLE 6	Finding the complex conjugate

Find the conjugate of each complex number.

a $z = -5 + 8i$ **b** $z = 4i(7 + 9i)$ **c** $z = 2i + \dfrac{3 - i^3}{4(3 + i)}$

Steps	Working
a	
b Expand and simplify.	
c 1 Simplify the algebraic fraction and write in the form $z = a + bi$.	
2 State the conjugate.	

MATCHED EXAMPLE 7 | Dividing complex numbers

Simplify each expression.

a $\dfrac{-1-2i}{2-i}$

b $\dfrac{-5}{7+4i}$

c $\dfrac{6+2i}{3i}$

Steps	Working
a 1 Multiply the numerator and denominator by the conjugate of the denominator, making the denominator a difference of 2 squares. **2** Simplify the numerator and denominator, making the denominator real.	
b 1 Multiply the numerator and denominator by the conjugate of the denominator, making the denominator a difference of 2 squares. **2** Simplify the numerator and denominator, and realise the denominator.	
c 1 Multiply the numerator and denominator by the conjugate of the denominator (multiplying by just i also works). **2** Simplify.	

SB

p. 399

If $z = -3 - 6i$ and $w = 5 + 4i$, evaluate each expression.

a $3z - 4w$ **b** zw **c** $\dfrac{z}{w}$

Steps	Working
a Substitute the complex numbers into the given expression and simplify.	
b Substitute the complex numbers into the given expression, expand and simplify.	
c 1 Substitute the complex numbers into the given expression.	
2 Write in the form $z = x + yi$.	

MATCHED EXAMPLE 9 | Factorising a quadratic expression in the set of complex numbers

Solve $2x^2 - 5x + 4 = 0$ and then factorise $2x^2 - 5x + 4$.

Steps	Working
1 Evaluate the discriminant to decide on the type of solution.	
2 Use the quadratic formula $x = \dfrac{-b \pm \sqrt{b^2 - 4ac}}{2a}$	
3 To factorise, write the solutions in the form $a + bi$.	
4 Write the expression in the form $(x - x_1)(x - x_2)$.	

Solve each quadratic equation and classify the roots as real, purely imaginary or complex.

a $3x^2 + 75 = 0$ **b** $2x^2 - 8x + 5 = 0$ **c** $3x^2 - 12x + 13 = 0$

p. 404

Steps	Working
a Make x^2 the subject and solve.	
b $2x^2 - 8x + 5 = 0$ does not factorise. Use the quadratic formula.	
c $3x^2 - 12x + 13 = 0$ does not factorise. Use the quadratic formula.	

MATCHED EXAMPLE 11 | Finding the quadratic equation given a complex root

If $z = 3 - 5i$ is a solution to $z^2 + bz + c = 0$, determine the values of b and c.

Steps	Working
1 Find the conjugate, which is the other solution.	
2 The sum of the roots is $-b$.	
3 The product of the roots is c.	
4 State the values found.	

Graph each complex number on an Argand diagram and find its modulus.

a $4 + 2i$ **b** $\overline{4 + 2i}$ **c** 4 **d** $2i$

Steps	Working
a **1** Graph $4 + 2i$ on an Argand diagram. **2** Use Pythagoras' theorem to find its modulus.	
b **1** Find the conjugate of $4 + 2i$. **2** Graph the point on an Argand diagram and use Pythagoras' theorem to calculate its modulus. Note that complex conjugates are reflections of each other in the real axis.	
c **1** Graph 4 on an Argand diagram. Purely real numbers lie on the real axis. **2** 4 is 4 units from the origin.	

d 1 Graph $2i$ on an Argand diagram.

Purely imaginary numbers lie on the imaginary axis.

2 $2i$ is 2 units from the origin.

9780170464109

MATCHED EXAMPLE 13 | Calculating the modulus

Find $|z|$ for each complex number.

a $z = 3 - \sqrt{7}i$ **b** $z = -4 - 4i$ **c** $z = \dfrac{5 - 10i}{\sqrt{5}}$

Steps	Working
a Use the formula.	
b Use the formula.	
c Separate into real and imaginary parts and use the formula.	

SB

Using CAS 3:
Modulus of a
complex number
p. 409

MATCHED EXAMPLE 14 | Rotation of a complex number

Find the complex number, w, formed when $z = 7 + 3i$ is rotated on the Argand diagram by each angle given.

a 90° anticlockwise **b** 90° clockwise

Steps	Working
a 1 Decide on the transformation involved.	
2 Identify the values of x and y and find w.	
b 1 Decide on the transformation involved.	
2 Identify the values of x and y and state w.	

| MATCHED EXAMPLE 15 | Converting from Cartesian to polar forms |

Write each complex number in polar form, expressing the angle in radians.

a $z = 3 + 3i$ **b** $z = 4 + 4\sqrt{3}i$ **c** $z = -6\sqrt{3} - 6i$

Steps	Working
a 1 Evaluate r.	
2 Find the angle θ.	
3 Write the complex number in polar form.	
b 1 Evaluate r.	
2 Find the principal argument.	
3 Write the complex number in polar form.	

c **1** Evaluate r.

$z = -6\sqrt{3} - 6i$

2 Find the principal argument.

3 Write the complex number in polar form.

MATCHED EXAMPLE 16 | Converting from polar to Cartesian forms

Write each complex number in Cartesian form.

a $z = 7\cos\left(-\dfrac{\pi}{2}\right) + 7\sin\left(-\dfrac{\pi}{2}\right)i$ **b** $z = \dfrac{1}{\sqrt{3}}\left(\cos\left(\dfrac{3\pi}{4}\right) + \sin\left(\dfrac{3\pi}{4}\right)i\right)$ **c** $z = 4\operatorname{cis}(60°)$

Steps	Working
a 1 Evaluate Re(z) and Im(z).	
2 Express z in Cartesian form.	
b 1 Evaluate Re(z) and Im(z).	
2 Express z in Cartesian form.	
c 1 Write the expression in expanded form.	
2 Evaluate Re(z) and Im(z).	
3 Express z in Cartesian form.	

SB

Using CAS 4:
Convert between
Cartesian and
polar form
p. 415

MATCHED EXAMPLE 17 | Multiplication and division of complex numbers

a Write $(5 + 3i)(7 - 6i)$ in the form $a + bi$.

b Given $z = 9\operatorname{cis}\left(\dfrac{3\pi}{4}\right)$ and $w = 3\operatorname{cis}\left(\dfrac{\pi}{2}\right)$, write $\dfrac{z}{w}$ in Cartesian form.

c Write z^2 in the form $a + bi$ given that $z = 3\operatorname{cis}(57°)$. Write a and b correct to two decimal places.

Steps	Working
a **1** Expand brackets.	
2 Simplify and express the answer in the required form.	
b **1** Express $\dfrac{z}{w}$ in polar form.	
2 Write in Cartesian form and simplify.	
c **1** Write z^2 in polar form.	
2 Write in Cartesian form, writing answers to the required accuracy.	

MATCHED EXAMPLE 18   Vertical and horizontal lines

a   Use $z = 6 + 7i$ and $z_0 = 3 + 4i$ to draw $\text{Re}(z - z_0)$.

b   Use $z = -4 + 2i$ and $z_0 = -8 + 6i$ to draw $\text{Im}(z - z_0)$.

Steps	Working
a **1** Write $z - z_0$ in the form $a + bi$.	
2 State $\text{Re}(z - z_0)$.	
3 Show $\text{Re}(z - z_0)$ on the complex number plane.	
b **1** Write $z - z_0$ in the form $a + bi$.	
2 State $\text{Im}(z - z_0)$.	
3 Show $\text{Im}(z - z_0)$ on the complex number plane.	

9

MATCHED EXAMPLE 19   Lines in the complex plane

Sketch each line in the complex plane.

For **b**, find the angle, to the nearest degree, the line makes with the positive real axis.

a $|z - 7| = |z + 3|$ **b** $|z - 2 + 2i| = |z - 5|$

Steps	Working
a 1 Decide if z_1 and z_2 are real.	
2 Use the appropriate form of the solution.	
3 Describe the solution.	
b 1 Substitute $z = x + yi$ into the equation.	
2 Write the expression in linear form.	
3 Sketch the straight line.	
4 Calculate the angle the line makes with the positive real axis.	

MATCHED EXAMPLE 20 | Sketching rays

a Use $z_0 = 3 + 2i$ to show $\arg(z - z_0) = 50°$.

b Use $z_0 = 4 + 5i$ to show $\arg(z - z_0) = -\dfrac{4\pi}{3}$.

Steps	Working
a Start at z_0 and draw a line that has been rotated anticlockwise by the given number of degrees.	
b Start at z_0 and draw a line that has been rotated clockwise by the given number of radians.	

| MATCHED EXAMPLE 21 | The Cartesian equation of a circle |

a Sketch the graph of $|z + 4 - 2i| = 4$.

b Express $|z + 4 - 2i| = 4$ in Cartesian form and state the radius of the circle and the coordinates of its centre.

Steps	Working
a 1 Determine the radius of the circle and the coordinates of its centre.	
2 Mark the points $(x_0 + r, y_0)$, $(x_0, y_0 - r)$, $(x_0 - r, y_0)$ and $(x_0, y_0 - r)$.	
3 Sketch the circle using the 4 points as a guide.	
b 1 Substitute for z and express in the right form.	
2 Find the magnitude.	
3 Express the equation in the form $(x - x_0)^2 + (y - y_0)^2 = r^2$	
4 State the radius and the centre.	

MATCHED EXAMPLE 22 | Converting the Cartesian equation of a circle to complex form

Write each equation in the form $|z - z_0| = r$.

a $(x + 3)^2 + (y - 8)^2 = 81$

b $x^2 + y^2 - 10x - 14y + 10 = 0$

Steps	Working
a 1 Find the centre and the radius.	
2 Write the equation in the form $|z - z_0| = r$.	
b 1 Complete the square. Apply complete the square to the x terms and to the y terms.	
2 Write the expression in the form $(x - x_0)^2 + (y - y_0)^2 = r^2$	
3 Find the centre and the radius.	
4 Write the equation in the form $|z - z_0| = r$.	

9

MATCHED EXAMPLE 23 | Sketching an ellipse with real foci

Draw the ellipse $|z - 6| + |z + 6| = 15$.

Steps	Working
1 State x_{width}.	
2 Identify the foci.	
3 Find the centre of the ellipse.	
4 Find y_{width}	
5 Use the centre as reference to locate the endpoints of x_{width} and y_{width}.	
6 Sketch the graph.	

9780170464109

MATCHED EXAMPLE 24 | Sketching an ellipse with imaginary foci

Draw the ellipse $|z - 2i| + |z + 2i| = 5$.

Steps	Working
1 State y_{width}.	
2 Identify the foci.	
3 Find the centre of the ellipse.	
4 Find x_{width}.	
5 Use the centre as reference to locate the endpoints of x_{width} and y_{width}.	
6 Sketch the graph.	

MATCHED EXAMPLE 25 | Sketching regions involving straight lines

Show each region in the complex plane.

a $\text{Re}(z + 4 - i) \geq 6$ **b** $|z - 4i| < |z + 2|$

Steps	Working
a 1 Describe the shape.	
2 Decide what region is to be shaded.	
3 Show the shaded region.	
b 1 Find the Cartesian equation.	
2 Draw the shape and shade in the required region.	

MATCHED EXAMPLE 26 | Sketching regions in circles and ellipses

Sketch each region in the complex plane.

a $|z - 4i| > 2$ **b** $|z + 3| + |z - 3| \leq 10$

Steps	Working
a 1 Describe the shape and the required region.	
2 Draw the shape and shade in the required region.	
b 1 Describe the shape and the region to be shaded.	
2 Draw the shape and shade in the required region.	

MATCHED EXAMPLE 27 | Sketching inequalities involving rays

Show each region in the complex plane.

a $\operatorname{Arg}(z-2)<\dfrac{2\pi}{5}$ **b** $-\dfrac{\pi}{2}<\operatorname{Arg}(z+1-5i)\le\dfrac{\pi}{3}$

Steps	Working
a 1 Draw the ray $\operatorname{Arg}(z-z_0)=\theta$.	
2 Work out the region to be shaded.	
b 1 Draw the two rays $\operatorname{Arg}(z-z_0)=\theta_1$ and $\operatorname{Arg}(z-z_0)=\theta_2$.	

2 Shade the appropriate region.

CHAPTER

10 VECTORS

MATCHED EXAMPLE 1 Vector notation

a On a set of coordinate axes, plot the points $P(-1,1), Q(-2,-3)$ and $R(2,-4)$.

b Sketch the vectors $\overrightarrow{PQ}$, $\overrightarrow{QR}$ and $\overrightarrow{RP}$.

Steps	Working
a Set up coordinate axes, and plot and label the given points.	
b Sketch the vectors $\overrightarrow{PQ}$, $\overrightarrow{QR}$ and $\overrightarrow{RP}$.	

On a set of coordinate axes, sketch the position vectors $\overrightarrow{OA}$ and $\overrightarrow{OB}$, where the points are $A(-1, 2)$ and $B(2, 3)$. Label the vectors $\underset{\sim}{a}$ and $\underset{\sim}{b}$.

SB
p. 441

Steps	Working
1 Set up coordinate axes, and plot and label the given points A and B.	
2 Sketch and label the vectors $\overrightarrow{OA}$ and $\overrightarrow{OB}$.	

MATCHED EXAMPLE 3 | Directed line segments

Draw a directed line segment corresponding to the column matrix $\begin{bmatrix} 9 \\ 3 \end{bmatrix}$.

Steps   **Working**

The vector $\begin{bmatrix} 9 \\ 3 \end{bmatrix}$ means across 9 and up 3 units.

Note this arrow goes from $(-1, 1) \rightarrow (8, 4)$.
It could have started at any point as long as
it goes across 9 and up 3 units.

MATCHED EXAMPLE 4 | Orthogonal unit vectors

a Write and sketch the directed line segment corresponding to the column matrix $\begin{bmatrix} 3 \\ 4 \end{bmatrix}$ in orthogonal unit form.

b Find the magnitude and direction of the column matrix $\begin{bmatrix} 3 \\ 4 \end{bmatrix}$.

Steps	Working
a The vector $\begin{bmatrix} 3 \\ 4 \end{bmatrix}$ means across 3 and up 4 units. Write in $\underset{\sim}{i}$ and $\underset{\sim}{j}$ form.	
b Find the magnitude and direction.	

MATCHED EXAMPLE 5 | Unit vectors

For the vector $\underset{\sim}{a} = \underset{\sim}{i} + 4\underset{\sim}{j}$, find the magnitude of $\underset{\sim}{a}$ and the unit vector in the direction of the vector $\underset{\sim}{a}$.

Steps	Working
1 Find the magnitude of the vector $\underset{\sim}{a}$.	
2 Find the unit vector, using $\hat{\underset{\sim}{a}} = \dfrac{\underset{\sim}{a}}{\lvert \underset{\sim}{a} \rvert}$	

For the vector $\underset{\sim}{a} = 3\underset{\sim}{i} + 4\underset{\sim}{j}$, find α and β, the angles in degrees correct to 2 decimal places between the vector $\underset{\sim}{a}$ and the x-axis and the y-axis, respectively.

SB
p. 445

Steps	Working
1 Find the magnitude of the vector $\underset{\sim}{a}$.	
2 The vector is in the form $\underset{\sim}{a} = a_1\underset{\sim}{i} + a_2\underset{\sim}{j}$, where $a_1 = 3$ and $a_2 = 4$. Use these values to find α and β.	

MATCHED EXAMPLE 7 | Operations with vectors

If $\underset{\sim}{u} = \begin{bmatrix} -2 \\ -1 \end{bmatrix}$ and $\underset{\sim}{v} = \begin{bmatrix} 3 \\ -5 \end{bmatrix}$, find

a $\underset{\sim}{u} + \underset{\sim}{v}$ **b** $\underset{\sim}{u} - \underset{\sim}{v}$ **c** $3\underset{\sim}{u} - 2\underset{\sim}{v}$

Steps	Working
a Evaluate $\underset{\sim}{u} + \underset{\sim}{v}$.	
b Evaluate $\underset{\sim}{u} - \underset{\sim}{v}$.	
c Evaluate $3\underset{\sim}{u} - 2\underset{\sim}{v}$.	

Find the values of p and q such that $2p\begin{bmatrix} 1 \\ -5 \end{bmatrix} - 3q\begin{bmatrix} -1 \\ 2 \end{bmatrix} = \begin{bmatrix} 7 \\ 12 \end{bmatrix}$.

SB p. 449

Steps	Working
1 Simplify the LHS using scalar multiplication.	
2 Subtract the column vectors.	
3 Equate the components of the vectors.	
4 Solve the simultaneous equations for p and q.	

TI-Nspire   **ClassPad**

| **5** Write the answer. | |

SB

p. 454

SB

Using CAS 2:
The scalar product
p. 454

SB

Using CAS 3:
Scalar product and
angle between
vectors
p. 455

MATCHED EXAMPLE 9 | Scalar product

If $\underset{\sim}{a} = -\underset{\sim}{i} + \underset{\sim}{j}$ and $\underset{\sim}{b} = 2\underset{\sim}{i} + 9\underset{\sim}{j}$, find $\underset{\sim}{a} \cdot \underset{\sim}{b}$.

Steps	Working
Use the component form formula $\underset{\sim}{a} \cdot \underset{\sim}{b} = a_1 b_1 + a_2 b_2.$	

MATCHED EXAMPLE 10 | Angles using dot product

If $|\underset{\sim}{a}| = 5$ and $|\underset{\sim}{b}| = 2$, and the angle between $\underset{\sim}{a}$ and $\underset{\sim}{b}$ is 30°, find $\underset{\sim}{a} \bullet \underset{\sim}{b}$.

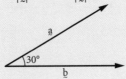

Steps	Working
Substitute into the formula $\underset{\sim}{a} \bullet \underset{\sim}{b} = |\underset{\sim}{a}||\underset{\sim}{b}|\cos(\theta)$.	

MATCHED EXAMPLE 11 Angle between 2 vectors

If $\underset{\sim}{a} = \underset{\sim}{i} + 3\underset{\sim}{j}$ and $\underset{\sim}{b} = 3\underset{\sim}{i} + 4\underset{\sim}{j}$, find the angle between $\underset{\sim}{a}$ and $\underset{\sim}{b}$ correct to the nearest degree.

Steps	Working
1 Find $\underset{\sim}{a} \bullet \underset{\sim}{b}$ and $|\underset{\sim}{a}||\underset{\sim}{b}|$.	
2 Substitute into the formula $\cos(\theta) = \dfrac{\underset{\sim}{a} \bullet \underset{\sim}{b}}{|\underset{\sim}{a}||\underset{\sim}{b}|}$, and solve for θ.	

MATCHED EXAMPLE 12 | Perpendicular vectors

a Show that the vectors $\underset{\sim}{a} = 3\underset{\sim}{i} - \underset{\sim}{j}$ and $\underset{\sim}{b} = 2\underset{\sim}{i} + 6\underset{\sim}{j}$ are perpendicular.

b For the perpendicular vectors $\underset{\sim}{u} = a\underset{\sim}{i} - \underset{\sim}{j}$ and $\underset{\sim}{v} = \sqrt{3}\underset{\sim}{i} + b\underset{\sim}{j}$, it is known that $|\underset{\sim}{v}| = \sqrt{84}$.
Find the value(s) of a and b.

SB
p. 458

Steps	Working
a **1** Find $\underset{\sim}{a} \bullet \underset{\sim}{b}$.	
2 Use the formula $\underset{\sim}{a} \bullet \underset{\sim}{b} = 0$.	
b **1** Find $\underset{\sim}{u} \bullet \underset{\sim}{v}$.	
2 Use the formula $\underset{\sim}{u} \bullet \underset{\sim}{v} = 0$ and $|\underset{\sim}{v}| = \sqrt{84}$.	
3 Solve the equations simultaneously to find a and b.	

MATCHED EXAMPLE 13   Parallel unit vectors

Find the unit vector parallel to the vector $\underset{\sim}{a} = 2\underset{\sim}{i} + 7\underset{\sim}{j}$.

Steps	Working
1 Find the length of $\underset{\sim}{a}$.	
2 Use the formula $\hat{\underset{\sim}{a}} = \dfrac{\underset{\sim}{a}}{\left|\underset{\sim}{a}\right|}$.	

MATCHED EXAMPLE 14 | Parallel vectors

Show that the vectors $\underset{\sim}{a} = \sqrt{3}\,\underset{\sim}{i} - 4\,\underset{\sim}{j}$ and $\underset{\sim}{b} = -2\sqrt{3}\,\underset{\sim}{i} + 8\,\underset{\sim}{j}$ are parallel in the opposite direction.

Steps	Working
1 Find $|\underset{\sim}{a}|$ and $|\underset{\sim}{b}|$	
2 Use the formula for parallel vectors $\underset{\sim}{a} \bullet \underset{\sim}{b} = \pm|\underset{\sim}{a}||\underset{\sim}{b}|$.	
3 Write the conclusion.	

10

MATCHED EXAMPLE 15 | Vector resolutes

Find the parallel and perpendicular components of $\underset{\sim}{a} = 3\underset{\sim}{i} + 4\underset{\sim}{j}$ in the direction of $\underset{\sim}{b} = \underset{\sim}{i} + \underset{\sim}{j}$.

Steps	Working
1 First, find $\hat{\underset{\sim}{b}}$.	
2 Find the vector projection of $\underset{\sim}{a}$ in the direction of $\underset{\sim}{b}$ using $(\underset{\sim}{a} \bullet \hat{\underset{\sim}{b}})\hat{\underset{\sim}{b}}$.	
3 Find the vector projection of $\underset{\sim}{a}$ perpendicular to $\underset{\sim}{b}$ using $\underset{\sim}{a} - (\underset{\sim}{a} \bullet \hat{\underset{\sim}{b}})\hat{\underset{\sim}{b}}$.	
4 Check the answer: the parallel and perpendicular components of $\underset{\sim}{a}$ should sum up to $\underset{\sim}{a}$.	

Find the vector projections of $\underset{\sim}{a} = 5\underset{\sim}{i} + \underset{\sim}{j} + 3\underset{\sim}{k}$ parallel and perpendicular to $\underset{\sim}{b} = 2\underset{\sim}{i} - \underset{\sim}{j} + \underset{\sim}{k}$.

SB
p. 463

Steps	Working
1 Find $\hat{\underset{\sim}{b}} = \dfrac{\underset{\sim}{b}}{\lvert \underset{\sim}{b} \rvert}$.	
2 Calculate the vector projection of $\underset{\sim}{a}$ on $\underset{\sim}{b}$.	
3 Find the vector projection of $\underset{\sim}{a}$ perpendicular to $\underset{\sim}{b}$.	

10

SB

p. 465

MATCHED EXAMPLE 17 | Statics

Two tow trucks are pulling a tractor stuck in mud by means of a steel cable, the first one with a force of 600 N at an angle of 47° with the horizontal and the second one with a force of 300 N at an angle of 63° with the horizontal. What is the total force acting on the tractor (to the nearest Newton), and in what direction is it acting (in degrees correct to one decimal place)?

Steps	**Working**
1 Draw a diagram to show the forces.	
2 Draw the forces head to tail, and complete the triangle with the resultant. Since $EB \parallel DC$, $\angle EBC = 180° - 63° = 117°$, so $\angle ABC = 47° + 117° = 164°$.	
3 Use the cosine rule on triangle ABC to find r.	
4 Use the sine rule on $\triangle ABC$ to find $\angle BCA$.	
5 Calculate the angle of the resultant with the horizontal.	
6 State the answer in an appropriate form.	

MATCHED EXAMPLE 18 | Vectors and velocity

The velocity of a sailing ship changes from 15 knots at 42° to 10 knots at a bearing of 130°.

What is the change in velocity? Give the magnitude and direction correct to one decimal place.

Steps	**Working**
1 Sketch a diagram.	
Let $\underset{\sim}{v}_1$ be the velocity of 15 knots at 42°.	
Let $\underset{\sim}{v}_2$ be the velocity of 10 knots at 130°.	
Mark $-\underset{\sim}{v}_1$ on the diagram.	
Rearrange $-\underset{\sim}{v}_1$ and $\underset{\sim}{v}_2$ in the diagram so that the change of velocity can be found.	
Find the angle between the vectors.	
The obtuse angle in the parallelogram is $42° + 50° = 92°$, so the acute angle is 88°.	
2 Draw a diagram to find $\underset{\sim}{v}_2 - \underset{\sim}{v}_1$.	
Use the cosine rule to find a, the magnitude of $\underset{\sim}{v}_2 - \underset{\sim}{v}_1$.	
Keep more decimal places in your calculator.	
3 Use the sine rule to find A, which will help us find the direction of $\underset{\sim}{v}_2 - \underset{\sim}{v}_1$.	
Keep more decimal places in your calculator.	
4 Calculate the bearing.	
5 Answer the question.	

MATCHED EXAMPLE 19 | Vector proof 1

In the given parallelogram, $\underset{\sim}{a} = \underset{\sim}{c}$, $\underset{\sim}{b} = \underset{\sim}{d}$, $\overrightarrow{AB} = \underset{\sim}{f}$ and $\overrightarrow{OC} = \underset{\sim}{e}$.

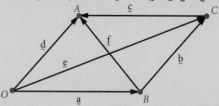

Which of the following statements is true?

A $\underset{\sim}{a} + \underset{\sim}{c} = \underset{\sim}{b} + \underset{\sim}{d}$

B $\underset{\sim}{e} \bullet \underset{\sim}{f} = 0$

C $\underset{\sim}{f} = \underset{\sim}{b} - \underset{\sim}{a}$

D $\underset{\sim}{d} = -\underset{\sim}{b}$

E $\underset{\sim}{e} = \underset{\sim}{a} + \underset{\sim}{d}$

Steps	Working
1 Consider what is given.	
2 Consider what we know.	
3 Consider what is clearly wrong.	
4 Consider the vector operations.	

Use vectors to prove that the diagonals of a rhombus intersect at right angles.

Steps	Working
1 Sketch a diagram to set up the proof.	
2 State the proof step by step.	
3 State the conclusion.	

SB

p. 470

CHAPTER

SAMPLING DISTRIBUTIONS

p. 482

MATCHED EXAMPLE 1	**Finding the probability distribution of a discrete random variable**

A box contains four bottles numbered $\{1, 2, 3, 4\}$. Two bottles are taken from the box, without replacement. If X represents the sum of the two numbers obtained, list the probability distribution of X.

Steps	**Working**

1 Draw a tree diagram to illustrate the problem.

2 Draw a probability distribution table.

X = the sum of the numbers and

$x \in \{3, 4, 5, 6, 7\}$.

3 Use the tree diagram to find the probabilities.

4 Write the probabilities in the table.

MATCHED EXAMPLE 2 | Finding the expected value of a discrete probability distribution

The probability distribution of a discrete random variable X is shown below.

Find the mean of X.

x	0	1	2	3
$\Pr(X = x)$	0.45	0.25	0.15	0.15

Steps	**Working**
Rewrite the table with rows as columns and add a column headed $x \times p(x)$ and calculate the product of the x values and their probabilities.	
The sum of the $x \times p(x)$ column is the expected value of X or $E(X)$.	

11

MATCHED EXAMPLE 3	Finding the variance and standard deviation of a discrete probability distribution

For the probability distribution below, find

a the expected value.

b the variance.

c the standard deviation, correct to three decimal places.

X	0	1	2	3	4
$p(x)$	0.3	0.3	0.2	0.1	0.1

Steps	Working
a Add two extra columns headed $x \times p(x)$ and $x^2 \times p(x)$ for calculating $E(X)$ and $E(X^2)$, respectively. Multiply x by $p(x)$, enter the results in the $x \times p(x)$ column and find the total. Multiply x by $x \times p(x)$, enter the results in the $x^2 \times p(x)$ column and find the total. The total of the $x \times p(x)$ column is $E(X)$.	
b The total of the $x^2 \times p(x)$ column is $E(X^2)$. Use the computational formula to find $\text{Var}(X)$.	
c Find the standard deviation using the formula $\text{SD}(X) = \sqrt{\text{Var}(X)}$.	

MATCHED EXAMPLE 4 Finding the mean and variance of 2X from a probability distribution

The probability distribution of a discrete random variable X is shown:

x	10	20	30	40
$p(x)$	0.4	0.2	0.1	0.3

a List the probability distribution of N where $N = 2X$.

b Hence, find

  i $E(N)$

  ii $\text{Var}(N)$

Steps	Working
a Complete a probability distribution table. Calculate the values of n by multiplying the x values by 2. The probabilities in the table are unchanged.	
b i Add two extra columns headed $n \times p(n)$ and $n^2 \times p(n)$. Complete the table calculations. Find $E(N)$ by finding the sum of the $n \times p(n)$ column.	
ii Find $E(N^2)$ by finding the sum of the $n^2 \times p(n)$ column. Use the computational formula $\text{Var}(X) = E(X^2) - (E(X))^2$ to find $\text{Var}(N)$.	

MATCHED EXAMPLE 5	Finding the mean and variance of 2X given the mean and variance of X

A discrete random variable Y has a mean of 8.5 and a variance of 9.

a Find the mean of $2Y$.

b Find the variance of $2Y$.

c Find the standard deviation of $2Y$.

Steps	Working
a Substitute the values of $E(Y)$ and k into $E(kY) = k \times E(Y)$.	
b Substitute the values of $\text{Var}(Y)$ and k into $\text{Var}(kY) = k^2 \times \text{Var}(Y)$.	
c Substitute in $\text{SD}(X) = \sqrt{\text{Var}(X)}$.	

A discrete random variable Y has a mean of 8.5 and a variance of 9.

MATCHED EXAMPLE 6	Finding the mean and variance of the sum of 2 identically distributed random variables 1

SB p. 490

The weather in England is unpredictable. On any day, it rains either 0, 1 or 2 times. The discrete random variable X representing the number of times it rains on any day is given by the distribution below.

x	0	1	2
$Pr(X = x)$	0.2	0.5	0.3

a Find the mean and variance of X.

b The weather reports are only collected every 2 days. Let T represent the total number of times it rained over two days.

Illustrate this situation with a tree diagram, if the number of times it rains on the first day is independent of the number of times it rains on the second day and then write the probability distribution for T.

c Find the mean and variance of T.

d Verify the results found in part **c** using the formula for the mean and variance of the sum of identical independent random variables.

Steps	Working
a Calculate the mean and variance of X from the probability distribution.	
b **1** Draw a tree diagram to represent the number of times it rained on two consecutive days.	

2 Calculate the probabilities of a total of 0, 1, 2, 3 and 4 times it rains.

List the probability distribution of T.

c Calculate the mean and variance of the total number of times it rains over two days, T.

d The total number of times it rains is the sum of two random values. $x_1 + x_2$

Substitute into the mean and variance formulas.

$E(x_1 + x_2) = 2 \times E(X)$

$\text{Var}(x_1 + x_2) = 2 \times \text{Var}(X)$

MATCHED EXAMPLE 7	Finding the mean and variance of the sum of 2 identically distributed random variables 2	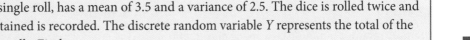

SB
p. 492

The probable outcomes of a fair dice are 1, 2, 3, 4, 5 and 6. The discrete random variable X, representing the outcome obtained on a single roll, has a mean of 3.5 and a variance of 2.5. The dice is rolled twice and the sum of the outcomes obtained is recorded. The discrete random variable Y represents the total of the outcomes obtained after two rolls. Find

a the mean of Y.

b the variance of Y.

c the standard deviation of Y. Round to three significant figures.

Steps	Working
a Substitute $E(X) = 3.5$ and $k = 2$ into $E(x_1 + x_2) = 2 \times E(X)$	
b Substitute $\text{Var}(X) = 2.5$ and $k = 2$ into $\text{Var}(x_1 + x_2) = 2 \times \text{Var}(X)$	
c Substitute in $\text{SD}(X) = \sqrt{\text{Var}(X)}$ and round to three significant figures.	

p. 496

Using CAS 2:
Generating random
numbers
p. 496

MATCHED EXAMPLE 8 Determining if a sample is random

A shopkeeper is conducting a survey on customer satisfaction. He asks a group of regulars to complete a survey he has designed. Will this result in a random sample?

Steps	Working
Simple random sampling ensures each element of the population has the same chance of being selected.	

MATCHED EXAMPLE 9 — Selecting a random sample

The table shows the height and weight values of 40 students in a classroom.

The heights are in centimetres (cm) and weights are in kilogram (kg).

Use a random number generator to select a random sample of eight items from this population.

No.	Height	Weight	No.	Height	Weight
1	100.2	25.6	21	114.4	32.2
2	112.0	37.1	22	99.6	25.7
3	105.5	25.0	23	95.7	26.4
4	100.8	28.3	24	109.8	31.1
5	114.6	33.3	25	111.5	33.9
6	118.5	31.2	26	103.6	29.6
7	109.7	30.0	27	118.3	35.71
8	104.4	22.8	28	109.1	30.2
9	115.0	29.8	29	112.5	27.6
10	119.1	28.4	30	96.6	22.9
11	113.3	26.6	31	113.9	34.8
12	111.9	31.5	32	100.9	27.3
13	116.2	34.5	33	114.1	32.8
14	105.2	28.8	34	118.6	37.8
15	103.2	26.3	35	119.4	35.4
16	107.9	30.7	36	104.8	28.3
17	100.6	24.9	37	106.7	27.1
18	98.9	24.0	38	97.3	25.6
19	96.0	23.6	39	115.9	33.4
20	120.3	38.9	40	97.5	28.1

Steps	Working
1 Generate 10 random numbers as shown in Using CAS 2. Include extra data to avoid duplicates.	
2 Identify the item numbers in the list that match the random numbers generated.	

MATCHED EXAMPLE 10 | Finding a sample proportion

A box contains five balls numbered 1, 2, 3, 4 and 5. Two balls are thrown eight times and the results below are obtained.

(2, 3) (4, 3) (2, 1) (5, 5) (5, 4) (4, 1) (1, 5) (3, 1)

Find the proportion of sums greater than or equal to 6.

Steps	Working
1 Count the number of items in the sample where the sum is greater than or equal to 6 and write the number of items in the sample.	
2 Substitute in the formula for sample proportion.	

MATCHED EXAMPLE 11 | Finding a sample mean

A survey was conducted in an apartment on the monthly expenditure for electricity by selecting 30 families in the apartment. The monthly expenditure for electricity, in dollars, are shown below.

140, 150, 145, 122, 118, 210, 221, 167, 198, 144, 136, 111, 100, 126, 147,

178, 124, 109, 116, 189, 174, 146, 131, 166, 101, 158, 129, 191, 155, 141

a Determine the sample mean $\bar{x}$, correct to one decimal place.

b Is the value $\bar{x}$ a statistic or a parameter?

Steps	Working
a 1 Find the sum of the expenses.	
2 Substitute into the formula $$\bar{x} = \frac{\sum x}{n}$$	
b 3 Determine if the mean relates to a sample or the population. Statistics are for samples. Parameters are for populations.	

p. 501

MATCHED EXAMPLE 12 | Listing an event and the event space

The spinner shown is spun twice.

a List the event space for this experiment.

b If *A* is the set of events where a 4 occurs on exactly one spin, list the elements of *A*.

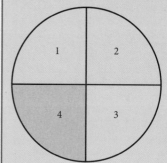

Steps	Working
a 1 Draw a lattice with 4 rows and 4 columns. Complete the cells in the matrix.	
2 Each cell represents an event or outcome. Write the outcomes as ordered pairs.	
b 3 Write the set that contains the events where a 4 occurs only once.	

9780170464109

MATCHED EXAMPLE 13 | Applying simulation methods

Two equally matched people, Brian and John, play badminton. A probability simulation with a spinner is divided into four equal sections numbered 1, 2, 3 and 4 is used to determine the results of three matches.

Let 1 and 3 represent John wins the set and 2 and 4 represent Brian wins the set.

The winner of a match is the first player to win 2 sets.

Use the simulations below to determine the match results.

1, 2, 4, 4, 3, 1, 4, 3, 3, 1, 1, 2, 3, 4, 1

Steps	Working
Draw a table showing the simulations, Brian or John wins and the progressive match score. Each time a player reaches 2 wins, they win the match and then a new match begins. Complete the table and convert the 1, 2, 3, 4 simulation to B wins or J wins. Complete until 3 match results are obtained.	

p. 507

MATCHED EXAMPLE 14 The sampling distribution of sample means

Samples of size 2 are selected, with replacement, from a shelf that contains toy cars numbered 1, 2, 3 and 4. Let the random variable $\overline{X}$ represent the means of the samples of size 2.

List the probability distribution of $\overline{X}$.

Steps	Working
1 List all possible samples of size 2 and calculate the mean of each sample.	
2 There is one of the 16 possible samples with a mean of 1. $\Pr(\overline{X}=1)=\dfrac{1}{16}$ Complete a probability distribution table that shows the different possible values of $\overline{X}$ and their probabilities.	

Using CAS 3: The distribution of sample means by simulation p. 509

9780170464109

MATCHED EXAMPLE 15 | Finding a probability from a simulation

A random sample is taken from a population that is normally distributed with a mean of 60 and a standard deviation of 8. Using simulation, 80 of these random samples is generated and the results are shown in the dot plot.

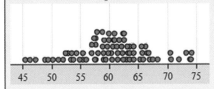

Find the probability a sample contains a mean greater than 70.

Steps	Working
Count the number of sample means greater than 70. The probability is a fraction of the total number of samples.	

MATCHED EXAMPLE 16 | Finding the average of the sample means by simulation

The intelligence quotient (IQ) of a random selection of students in a college is normally distributed with a mean of 120 and a standard deviation of 4.

a Simulate 40 samples of size 30 and calculate the sample means for each sample.

b Find the average of the sample means for the samples of size 30.

c Display the results as a dot plot or histogram.

Steps	Working

a Generate the means of the samples using CAS randNorm command.

TI-Nspire   ClassPad

b Calculate one-variable statistics for the data in column A.

Mean of $\bar{x} = 119.975$

TI-Nspire

Calculate one-variable statistics for the data in column A.

Mean of $\bar{x} = 120.4515$

ClassPad

c Graph the data.

TI-Nspire   ClassPad

MATCHED EXAMPLE 17	Comparing the standard deviation of the sample means for 2 different sample sizes

Random samples of size n are taken from a population that is normally distributed with a mean of 80 and a standard deviation of 12. Complete 100 simulations for sample sizes of 40 and 110 and determine the mean and standard deviation of the sample means.

Summarise this information in the table below.

Sample size	Mean of $\bar{x}$	Standard deviation of $\bar{x}$
40		
110		

Steps	Working

Generate the means of the samples by CAS using the randNorm command.

Cell formula

Use $n = 40$ and $n = 110$ in the formula.

p. 514

MATCHED EXAMPLE 18 | Finding the standard error of a sample mean

A random sample of 150 rowers is taken at a rowing championship. The weight of the rowers is normally distributed with a mean weight of 110 kg and a standard deviation of 6 kg. Find the standard error in the sample mean. Round to three significant figures.

Steps	Working
Use the formula $SE(\bar{x}) = \dfrac{\sigma}{\sqrt{n}}$. Round to three significant figures.	

TRANSFORMATIONS AND MATRICES

CHAPTER

12

MATCHED EXAMPLE 1	Translation of a shape

SB

p. 527

Find the image of the rectangle $P(-1, 2)$ $Q(1, 1)$ $R(3, 5)$ $S(1, 6)$ under the translation $P(x, y) \rightarrow P'(x - 4, y + 5)$. What is the shape of the image?

Steps	Working
1 Find the image points by substituting the coordinates into the transformation.	
2 Draw the original and image.	
3 Write the answer.	

SB

p. 528

MATCHED EXAMPLE 2	Linear transformation of a shape

Find the image of the isosceles triangle $A\,(2, 1)$ $B\,(6, 4)$ $C\,(-1, 5)$ under the transformation $P\,(x, y) \rightarrow P'\,(4x + 2y, x - 2y)$. What is the shape of the image?

Steps	Working
1 Find the image points by substituting the coordinates into the transformation.	
2 Draw the image.	
3 It looks like a scalene triangle. Check side lengths to confirm.	
4 Write the answer.	

SB

Using CAS 1: Linear transformation of a shape p. 529

MATCHED EXAMPLE 3 Translations using matrices

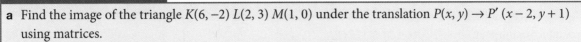

a Find the image of the triangle $K(6, -2)$ $L(2, 3)$ $M(1, 0)$ under the translation $P(x, y) \rightarrow P'(x - 2, y + 1)$ using matrices.

b What is the image of the point $(-2, 3)$ after translation 2 up and 5 to the right, followed by translation 4 up and 3 to the left?

Steps	Working
a 1 Find the image points by adding matrices.	
2 Write the answer.	
b 1 Add the translations.	
2 Find the image point.	
3 Write the answer.	

MATCHED EXAMPLE 4 | Linear transformation using a matrix

Apply matrix multiplication to the image of the isosceles triangle A (2, 1) B (6, 4) C (−1, 5) under the transformation $P(x, y) \rightarrow P'(4x + 2y, x − 2y)$.

Steps	Working
1 Write the matrix.	
2 Multiply the matrix by A (2, 1) as a column vector.	
3 Multiply the matrix by B (6,4) as a column vector.	
4 Multiply the matrix by C (−1,5) as a column vector.	
5 Write the answer.	

MATCHED EXAMPLE 5 | Dilation of a function

Find the image of the curve $y = x^2 + 2$ under dilation by a factor of 2 parallel to the x-axis. Sketch the curves and comment on the transformation.

Steps	Working
1 Write the transformation.	
2 Express x and y in terms of x' and y'.	
3 Substitute into the equation of the line and express in the normal form.	
4 Sketch the parabolas.	
5 Write the answer.	

MATCHED EXAMPLE 6 | Rotation of a shape

Find the image of the triangle A (2, 2) B (2, −2) C (4, 0) after rotation through an angle of −135° around the origin.

Steps	Working
1 Write the transformation.	
2 Substitute values.	
3 Apply the transformation to each point.	
4 Simplify.	
5 Write the answer.	

MATCHED EXAMPLE 7 | Reflection of a function

Find the image of the line $y = x - 1$ after reflection in the line through the origin with inclination 120°.

p. 536

Steps	Working
1 Write the transformation.	
2 Simplify.	
3 Express x and y in terms of x' and y' and eliminate fractions.	
4 Eliminate y to find x.	
5 Eliminate x to find y.	
6 Substitute in $y = x - 1$ and simplify.	
7 Rationalise the denominators.	
8 Write the answer.	

MATCHED EXAMPLE 8 | Inverse of a dilation

a Find the inverse of a dilation by the factor 1.25 from the x-axis.

b Confirm they are inverses by applying them in succession to a general point

Steps	Working
a 1 Find the reciprocal of 1.25	
2 Write the inverse.	
b 1 Write both transformations.	
2 Apply the first to (a, b).	
3 Now apply the second.	
4 State the conclusion.	

MATCHED EXAMPLE 9 | Inverse of a linear transformation

Find the inverse of the transformation $T: (x, y) \rightarrow (x + 2y, 3x + 4y)$ and confirm that it is the inverse by using the point $(4, -2)$.

Steps	Working
1 Find the value of the determinant.	
2 Write the rule for the inverse.	
3 Substitute in the values.	
4 Simplify.	
5 Apply the transformation to $(4, -2)$.	
6 Apply the inverse to $(0, 4)$.	
7 Write the answer.	

SB

Using CAS 4:
Inverse
transformations
p. 541

MATCHED EXAMPLE 10 | Using a composition of 2 transformations

$T: (x, y) \rightarrow (x + 2y, 3x + 5y)$, $S: (x, y) \rightarrow (x, 2y)$ and $R: (x, y) \rightarrow (x + 3, y - 2)$. Find:

a $T \circ S\,(4, -3)$ **b** $S \circ T\,(4, -3)$ **c** $T \circ R\,(4, -3)$ **d** $R \circ T(4, -3)$

Steps	Working
a 1 Do *S* first.	
2 Now do *T* on the result.	
3 Write the final answer.	
b 1 Do *T* first.	
2 Now do *S* on the result.	
3 Write the final answer.	
c 1 Do *R* first.	
2 Now do *T* on the result.	
3 Write the result.	
d Do *T* first. Now do *R* on the result. Write the result.	

MATCHED EXAMPLE 11 | Rule for a composition

$T: (x, y) \rightarrow (x + 2y, 3x + 5y)$ and $S: (x, y) \rightarrow (2x - 3y, x + 2y)$. Find $T \circ S$.

p. 545

Steps	Working
1 Do S first on a general point.	
2 Now do T on the result.	
3 Simplify.	
4 Write the overall answer.	

12

SB

p. 546

MATCHED EXAMPLE 12 | Using matrices for a composition

$T: (x, y) \rightarrow (x + 2y, 3x + 5y)$ and $S: (x, y) \rightarrow (2x - 3y, x + 2y)$. Use matrices to find $T \circ S$.

Steps	Working
1 Write S as a matrix.	
2 Write T as a matrix.	
3 Find $T \circ S$ as the matrix product.	
4 Write the overall answer.	

SB

Using CAS 5:
Composition using
matrices
p. 546

MATCHED EXAMPLE 13 | Linear transformation of a straight line

Show that the image of a straight line with slope m is another straight line with slope $\dfrac{r+ms}{p-mq}$ under the non-singular linear transformation $T:(x,y)\rightarrow(px-qy,rx+sy)$

Steps	Working
1 Write the equation of a straight line L.	
2 Find the image of (x, y) under T.	
3 Write x' and y' in terms of x and y.	
4 Express x and y in terms of x' and y'.	
5 Substitute in the equation of L.	
6 Simplify and express y' in terms of x'.	
7 Write the conclusion.	

MATCHED EXAMPLE 14 | Areas under linear transformations

Show that $A\,(1, 1)\;B(2, -1)\;C(6, 1)\;D(5, 3)$ is a rectangle and find the area of its image under the transformation $T: (x, y) \rightarrow (x + y, 2x - y)$.

Steps	Working
1 Check that $ABCD$ is a rectangle. Find the gradients of the sides.	
2 Write the result.	
3 Find the lengths of perpendicular sides.	
4 Find the area.	
5 Find the determinant of T.	
6 Find the area of the image.	
7 Write the answer.	

MATCHED EXAMPLE 15 | Transformation that is its own inverse

Show that $T: (x, y) \to (4x + 5y, -3x - 4y)$ is its own inverse.

Steps	Working
1 Write T as a matrix.	
2 Multiply it by itself.	
3 Write the conclusion.	

p. 552

MATCHED EXAMPLE 16 | Proving a property of a transformation

Show that the image of an isosceles triangle with its base parallel to the y-axis is also isosceles under a dilation from the y-axis.

Steps	Working
1 Choose a general isosceles triangle, $\triangle ABC$.	
2 Check that it is isosceles.	
3 Choose a general dilation from the y-axis.	
4 Apply the transformation to $\triangle ABC$.	
5 Find the lengths of $A'C'$ and $B'C'$.	
6 Write the conclusion.	

9780170464109

MATCHED EXAMPLE 17 | Non-commutativity

Show that composition of a dilation and a rotation is not commutative.

Steps	Working
1 Determine what you need to do.	
2 Try a dilation and a reflection.	
3 Write the matrices.	
4 Find $D \circ R$.	
5 Find $R \circ D$.	
6 State the result.	

SB

p. 554

12

MATCHED EXAMPLE 18 | Equivalence of transformations

Find the composition of reflection in the line $y = x \tan(30°)$ and then in $y = 3x$ and show that it is equivalent to a rotation.

Steps	Working
1 Write the matrix for the first reflection.	
2 Find p and q for the second reflection.	
3 Write the matrix for the second reflection.	
4 Find the composition.	
5 Simplify.	
6 Check that the values could be sine and cosine of the same angle.	
7 State the values of $\sin(\theta)$ and $\cos(\theta)$.	
8 Write the product matrix in terms of $\sin(\theta)$ and $\cos(\theta)$, taking into account the signs of $\sin(\theta)$ and $\cos(\theta)$.	
9 Write the answer.	

MATCHED EXAMPLE 19 | Equivalent transformations

Use equivalent rotations to show that $\cos(a+b)=\cos(a)\cos(b)-\sin(a)\sin(b)$

Steps	Working
1 Write the matrices for rotations around the origin through angles a and b.	
2 Find the composition of rotations around the origin through angles a and b.	
3 Simplify.	
4 Write the equivalence.	
5 Write the matrix for rotation around the origin through angle $a + b$.	
6 Equate the cosine terms.	

Answers

CHAPTER 1

MATCHED EXAMPLE 1

$t_n = t_{n-1} \times 2$

MATCHED EXAMPLE 2

0, 2, 6, 12, 20

MATCHED EXAMPLE 3

a The sequence is arithmetic.

The next 4 terms are 14, 17, 20, 23.

b The sequence is arithmetic.

The next 4 terms are

$2 + 4\sqrt{3}, 2 + 5\sqrt{3},$

$2 + 6\sqrt{3}, 2 + 7\sqrt{3}$

c The sequence is NOT arithmetic.

MATCHED EXAMPLE 4

The 17th term of the sequence is 82.

MATCHED EXAMPLE 5

The 34th term in the sequence is 138.

MATCHED EXAMPLE 6

3, 6, 9, 12, 15

MATCHED EXAMPLE 7

$S_{12} = 258$

MATCHED EXAMPLE 8

There are 10 terms in the series.

MATCHED EXAMPLE 9

969

MATCHED EXAMPLE 10

a $r = 4$

b $r = \dfrac{1}{3}$

MATCHED EXAMPLE 11

The eleventh term is 5 242 880.

MATCHED EXAMPLE 12

$\dfrac{1}{244\,140\,625}$ is the 14th term of the geometric sequence.

MATCHED EXAMPLE 13

$t_9 = 45\,927$

MATCHED EXAMPLE 14

No

MATCHED EXAMPLE 15

$\dfrac{1\,301\,742}{1\,953\,125}$

MATCHED EXAMPLE 16

There are 12 terms in the series.

MATCHED EXAMPLE 17

The first term of the geometric series is 1.

MATCHED EXAMPLE 18

The limiting sum is $\dfrac{15}{4}$.

MATCHED EXAMPLE 19

$0.2\overset{..}{1} = \dfrac{7}{33}$

MATCHED EXAMPLE 20

$t_n = n^3$

MATCHED EXAMPLE 21

$1, 2\dfrac{1}{3}, 2\dfrac{7}{9}, 2\dfrac{25}{27}, 2\dfrac{79}{81}$

MATCHED EXAMPLE 22

The general term is $t_n = \dfrac{1}{2}(5^n + 1)$.

MATCHED EXAMPLE 23

$t_n = \dfrac{3}{2} \times 5^{n-1} - \dfrac{1}{2}$

MATCHED EXAMPLE 24

The amount after 5 years is given by $\$5000 \times 1.075^5$.

MATCHED EXAMPLE 25

The value after 9 years and 4 months is $\$6500 \times 1.0025^{112}$.

MATCHED EXAMPLE 26

a $A_{n+1} = 1.07\,A_n + 3000$, where $A_1 = 3000$

b The value after n years is $3000\left(\dfrac{1.07^n - 1}{0.07}\right)$

c The value of Mark's investment when he turns

35 would be given by $3000\left(\dfrac{1.07^{13} - 1}{0.07}\right)$

MATCHED EXAMPLE 27

a $A_{n+1} = 1.105\,A_n - 85\,000$, where $A_1 = 750\,000$

b The amount to pay after n years is

$750\,000(1.105^n) - 85\,000\left(\dfrac{1.105^n - 1}{0.105}\right)$.

c The amount to pay after 7 years is

$750\,000(1.105^7) - 85\,000\left(\dfrac{1.105^7 - 1}{0.105}\right)$

MATCHED EXAMPLE 28

The payment required is $450\,000(1.065^{25})\left(\dfrac{0.065}{1.065^{25}-1}\right)$.

MATCHED EXAMPLE 29

The population in 2030 will be $1.02^8 \times 2\,500\,000$.

CHAPTER 2

MATCHED EXAMPLE 1

a $1.\dot{4}=1\dfrac{4}{9}$

b $0.\dot{5}\dot{3}=\dfrac{53}{99}$

c $2.1\dot{2}5\dot{6}=2\dfrac{251}{1998}$

MATCHED EXAMPLE 2

a $\sqrt{289}\in N$

b $\{x:x^2<9\}\subset R$

c $27^{\frac{1}{3}}+0.\dot{3}\in Q$

d $\{x:-7<x<5\}\subset R$

e $(2+3i)+(1-5i)\in C$

MATCHED EXAMPLE 3

a The set of rational numbers is closed under subtraction.

b The set of rational numbers is closed under multiplication.

c The set of rational numbers is not closed under division.

MATCHED EXAMPLE 4

Proof: See worked solutions

MATCHED EXAMPLE 5

a $A\cup B=\{0,1,2,3,4,5,7,9\}$

b $B\cap C=\{\varnothing\}$

c $\left(A\cap(B\cup C)\right)'=\{-2,-1,0,1,2,3,5,6,7,8,9,10,11,12\}$

d $n(A\cup C)'=10$

MATCHED EXAMPLE 6

a
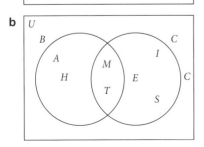

b

MATCHED EXAMPLE 7

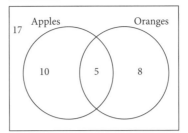

MATCHED EXAMPLE 8

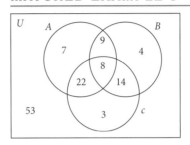

MATCHED EXAMPLE 9

a
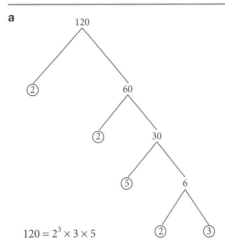

$120=2^3\times 3\times 5$

b The factors of 120 that are greater than 1 are 2, 3, 4, 5, 6, 8, 10, 12, 15, 20, 24, 30, 40, 60 and 120.

MATCHED EXAMPLE 10

a $2\sqrt{155}$ **b** 27

MATCHED EXAMPLE 11

Proof: See worked solutions

MATCHED EXAMPLE 12

Proof: See worked solutions

MATCHED EXAMPLE 13

a **1** 1 is neither prime nor composite.

    2 Therefore, not all natural numbers are either prime or composite.

b **1** For $n=3$, $n^2=3^2$, that is, 9 is odd.

    2 If n is an integer, then n^2 is not always even.

c **1** 2 is an even integer that is not divisible by 4.

    2 Therefore, not all even integers are divisible by 4.

MATCHED EXAMPLES 14–22

Proofs: See worked solutions

CHAPTER 3

MATCHED EXAMPLE 1

a Vertices *A, B* and *F* are adjacent to vertex *E*.

b Vertex *F* has a loop.

c Vertex *C* is isolated.

d Vertices *A* and *D* have multiple edges.

MATCHED EXAMPLE 2

a Vertices = {*A, B, C, D, E, F*}

b Edges = {*AB, AD, AD, AE, BE, EF, FF*}

MATCHED EXAMPLE 3

$$
\begin{array}{c@{}c}
 & \begin{array}{cccc} A & B & C & D \end{array} \\
\begin{array}{c} A \\ B \\ C \\ D \end{array} &
\left[\begin{array}{cccc}
0 & 1 & 1 & 0 \\
1 & 0 & 1 & 1 \\
1 & 1 & 1 & 0 \\
0 & 1 & 0 & 0
\end{array} \right]
\end{array}
$$

MATCHED EXAMPLE 4

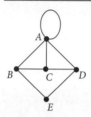

MATCHED EXAMPLE 5

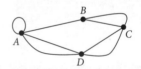

MATCHED EXAMPLE 6

deg (*A*) = 3

deg (*B*) = 2

deg (*C*) = 1

deg (*D*) = 4

MATCHED EXAMPLE 7

9

MATCHED EXAMPLE 8

a The graph is simple, connected and planar.

b The graph is simple.

c The graph is simple, connected, non-planar and bipartite.

MATCHED EXAMPLE 9

28

MATCHED EXAMPLE 10

10 vertices and 45 edges

MATCHED EXAMPLE 11

Option E

MATCHED EXAMPLE 12

10

MATCHED EXAMPLE 13

Option D

MATCHED EXAMPLE 14

a This graph is planar.

b This graph is not planar.

MATCHED EXAMPLE 15

There are two faces.

MATCHED EXAMPLE 16

Four faces

MATCHED EXAMPLE 17

Option B

MATCHED EXAMPLE 18

The maximum number of edges is 4.

MATCHED EXAMPLE 19

a This graph is not a tree.

b This graph is a tree.

c This graph is not a tree.

MATCHED EXAMPLE 20

There are eight vertices.

MATCHED EXAMPLE 21

a *A–B–C–E–D–B* is an open walk and a trail.

b *E–B–D–E* is a closed walk, a circuit and a cycle.

c *A–B–C–E–D* is an open walk, a trail and a path.

d *D–B–A–E–D* is a closed walk, a circuit and a cycle.

MATCHED EXAMPLE 22

a The Euler path finishes at *B*.

b Eulerian trails

   A–B–C–D–E–B

   A–B–E–D–C–B

MATCHED EXAMPLE 23

a The Hamiltonian path is *A–B–C–D–E–F*.

b The Hamiltonian cycle is *A–B–C–D–E–F–A*.

CHAPTER 4

MATCHED EXAMPLE 1

a Sentences 1, 5 and 7 are statements that are not used to form conclusions.

b Sentences 5 and 7 are premises because they are used to draw conclusion 8.

c Sentences 2, 3, 4 and 6 cannot be said to be true or false.

MATCHED EXAMPLE 2

a Zuri is a singer and Ella is an actor.

b If Ella is an actor, then Diana is not a dancer.

c If Zuri is not a singer, then Karla is a rapper.

d If Ella is not an actor, then either Diana is a dancer or Karla is a rapper.

MATCHED EXAMPLE 3

A	B	$\neg A \wedge B$
T	T	F
T	F	F
F	T	T
F	F	F

MATCHED EXAMPLE 4

A	B	$\neg(\neg A \vee B)$	$A \wedge \neg B$
T	T	F	F
T	F	T	T
F	T	F	F
F	F	F	F

MATCHED EXAMPLE 5

A	B	$A \rightarrow B$
T	T	T
T	F	F
F	T	T
F	F	T

MATCHED EXAMPLE 6

A	B	$A \rightarrow B$	$B \rightarrow A$	$(A \rightarrow B) \vee (B \rightarrow A)$
T	T	T	T	T
T	F	F	T	T
F	T	T	F	T
F	F	T	T	T

MATCHED EXAMPLE 7

A	B	$B \wedge \neg(A \rightarrow B)$
T	T	F
T	F	F
F	T	F
F	F	F

MATCHED EXAMPLE 8

a 101111_2

b 38_{10}

MATCHED EXAMPLE 9

a $100_2 + 110_2 = 1010_2$

b $1010_2 - 1001_2 = 1_2$

MATCHED EXAMPLE 10

a $101_2 \times 10_2 = 1010_2$

b $5 \times 2 = 10$

MATCHED EXAMPLE 11

$11100_2 \div 10_2 = 01110_2$ (or 1110_2)

MATCHED EXAMPLE 12

a $10_{10} + 0.375_{10} = 10.375_{10}$

b $12.25_{10} = 1100.01_2$

c $0.6_{10} = 10011001...._2$

MATCHED EXAMPLE 13

$(X + Y)(\overline{X} + Y) = Y$

MATCHED EXAMPLE 14

$(X\overline{Z} + \overline{Y}X) \bullet (X\overline{Z} + \overline{X}) = X\overline{Z}$

MATCHED EXAMPLE 15

$\overline{A}\overline{B}C + \overline{A}BC + ABC = C\overline{A} + CB$

MATCHED EXAMPLE 16

$(A + \overline{B})(A + C) = \overline{B}C + A$

MATCHED EXAMPLE 17

a (red OR green) AND marbles

b 'sports' NOT football

c (singing AND dancing) NOT (classical OR salsa)

MATCHED EXAMPLE 18

a

	A	B	C	D
1	Cairo	Cairo	1	
2	Madrid	Athens	0	
3	Paris	Paris	1	
4	Berlin	Rome	0	
5	Santiago	Santiago's	0	

b

	A	B	C	D
1	Cairo	Cairo	1	0
2	Madrid	Athens	0	1
3	Paris	Paris	1	0
4	Berlin	Rome	0	1
5	Santiago	Santiago's	0	1

MATCHED EXAMPLE 19

The SOP is $\overline{A}\overline{B} + AD + \overline{B}C + \overline{A}\overline{D}$.

MATCHED EXAMPLE 20

The SOP is $\overline{B} + \overline{D} + \overline{A}C$.

MATCHED EXAMPLE 21

a The SOP is B.

b The negation of B is $\overline{B}$.

MATCHED EXAMPLE 22

A	B	C	OR $A + B + C$
0	0	0	0
0	0	1	1
0	1	0	1
1	0	0	1
0	1	1	1
1	0	1	1
1	1	0	1
1	1	1	1

MATCHED EXAMPLE 23

$\overline{A}\overline{B} + BC(B + C)$ or $\overline{A} + \overline{B} + C$

MATCHED EXAMPLE 24

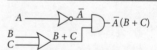

MATCHED EXAMPLE 25

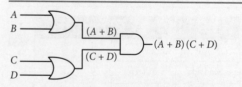

MATCHED EXAMPLE 26

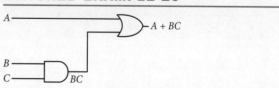

MATCHED EXAMPLE 27

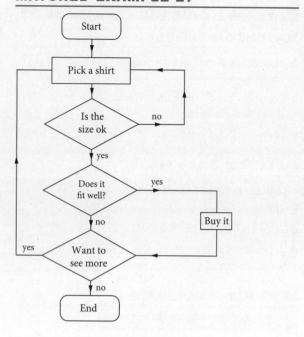

MATCHED EXAMPLE 28

a sum = 1 + 2 = 3

count = 2 + 1 = 3

sum = 3 + 3 = 6

count = 3 + 1 = 4

sum = 6 + 4 = 10

count = 4 + 1 = 5

sum = 10 + 5 = 15

count = 5 + 1 = 6

sum = 15 + 6 = 21

count = 6 + 1 = 7

sum = 21 + 7 = 28

count = 7 + 1 = 8

sum = 28 + 8 = 36

count = 8 + 1 = 9

sum = 36 + 9 = 45

count = 9 + 1 = 10

sum = 45 + 10 = 55

count = 10 + 1 = 11

b The algorithm finds the sum of the first 10 natural numbers.

MATCHED EXAMPLE 29

a Numbers from 1 to 20 with remainder 3 after division by 5 are.

b 3, 8, 13 and 18

MATCHED EXAMPLE 30

There are 30 numbers to read, one at a time.

INTEGER array[30]

current sum ← 0

FOR count = 1 TO 30

INPUT next number from list

procedure

ENDFOR

A number is even if the remainder after division by 2 is 0.

current number from the list ← 1

FOR count = 1 TO 30

   number ← RAND(50)

ENDFOR

current sum ← 0

FOR count = 1 TO 30

   INPUT next number from list

   current sum ← current sum + number from list

ENDFOR

PRINT "The sum of the 30 numbers is"

PRINT current sum

MATCHED EXAMPLE 31

There are six random numbers to generate.

current sum ← 0

FOR count = 1 TO 6

   procedure

ENDFOR

A number is even if the remainder after division by 2 is 0.

current sum ← 0

FOR count = 1 TO 6

REPEAT

   random number ← RAND(34)

   remainder ← random number MOD 2

UNTIL remainder = 0

PRINT random number

ENDFOR

current sum ← 0

FOR count = 1 TO 6

  REPEAT

   random number ← RAND(34)

   remainder ← random number MOD 2

   UNTIL remainder = 0

PRINT random number

current product ← current sum + random number

ENDFOR

 PRINT current sum

MATCHED EXAMPLE 32

INTEGER number[50]

INTEGER reverse[50]

current count ← 0

FOR loop = 1 TO 50

REPEAT

      array[loop]← RAND(50)

      UNTIL array[loop] MOD2 = 0

      count ← count + 1

      ENDFOR

      FOR loop = 1 TO count

      reverse[loop] ← array[count + 1 − loop]

ENDFOR

CHAPTER 5

MATCHED EXAMPLE 1

a P is a 2×3 matrix, Q is a row matrix of dimension 1×3 or just 3 and R is a square matrix with dimension 2.

b $p_{23} = -1, p_{12} = -6, q_3 = 8$ and $r_{21} = 3$

MATCHED EXAMPLE 2

$$3P = \begin{bmatrix} 12 & -9 & 6 \\ 15 & 3 & -21 \end{bmatrix}$$

$$-Q = \begin{bmatrix} 10 & 2 & 0 \\ -12 & -5 & -8 \end{bmatrix}$$

$$P + R = \begin{bmatrix} 4 & 5 & -5 \\ 13 & 0 & 4 \end{bmatrix}$$

$$Q + (-Q) = 0$$

$$Q - R = \begin{bmatrix} -10 & -10 & 7 \\ 4 & 6 & -3 \end{bmatrix}$$

$$4P + 5R = \begin{bmatrix} 16 & 28 & -27 \\ 60 & -1 & 27 \end{bmatrix}$$

MATCHED EXAMPLE 3

$$PQ = \begin{bmatrix} -8 & 17 & 18 \\ 3 & -4 & -8 \\ -4 & 18 & 4 \end{bmatrix}$$

$$QP = \begin{bmatrix} 3 & 13 \\ -14 & -11 \end{bmatrix}$$

MATCHED EXAMPLE 4

a $A(B+C) = \begin{bmatrix} 0 \\ 20 \end{bmatrix}$

b $AC + B = \begin{bmatrix} -2 \\ 22 \end{bmatrix}$

MATCHED EXAMPLE 5

a $14A - 5B$

b $-2(29F + 2C)$

c $9XY - 3YX$

d $11PQ + 6P - 13Q + 2QP$

MATCHED EXAMPLE 6

a $3MN + 5ML$

b $28AB - 20A$

c $3X^2 - 6XY - 4YX + 8Y^2$

MATCHED EXAMPLE 7

a $2X(Y - 2T)Z$

b $(7A + 2I)B$

c $2P(4P + Q) - (2P + Q)Q$

d $(4A - 5B)(4A + 5B)$

MATCHED EXAMPLE 8

$$\begin{bmatrix} \dfrac{1}{7} & 0 \\ 0 & \dfrac{1}{4} \end{bmatrix}$$

MATCHED EXAMPLE 9

a A is invertible.

b B is singular.

c C is non-singular.

MATCHED EXAMPLE 10

$$P^{-1} = \begin{bmatrix} 0.25 & 0.75 \\ 0 & 1 \end{bmatrix}$$

Q is singular and an inverse doesn't exist.

MATCHED EXAMPLE 11

a A is non-singular.

b B is invertible.

c C is invertible.

MATCHED EXAMPLE 12

a $|A| = -13$

$$A^{-1} = \begin{bmatrix} -\dfrac{5}{13} & \dfrac{3}{13} \\ \dfrac{6}{13} & -\dfrac{1}{13} \end{bmatrix}$$

b The determinant of $\begin{bmatrix} 2 & -4 \\ 1 & -2 \end{bmatrix}$ is 0, so it is singular and there is no inverse.

c $|C| = 15$

$$C^{-1} = \begin{bmatrix} \dfrac{1}{3} & 0 \\ \dfrac{4}{15} & \dfrac{1}{5} \end{bmatrix}$$

MATCHED EXAMPLE 13

a $|A| = -12$

b $|B| = -45$

MATCHED EXAMPLE 14

$|G| = -55$

MATCHED EXAMPLE 15

a $X = \begin{bmatrix} 0.6 & 1.4 \\ 1 & 1.4 \end{bmatrix}$

b $Y = \begin{bmatrix} \dfrac{28}{9} & \dfrac{2}{9} & 1 \\ \dfrac{25}{9} & \dfrac{13}{9} & \dfrac{14}{9} \end{bmatrix}$

MATCHED EXAMPLE 16

a $X = \begin{bmatrix} -4.5 & 3 \\ -3.5 & -1 \end{bmatrix}$

b $X = \begin{bmatrix} -\dfrac{7}{5} & 4 \\ -\dfrac{14}{5} & 6 \end{bmatrix}$

c $X = \begin{bmatrix} -4 & -4 \\ -0.166\ldots & -1.333\ldots \end{bmatrix}$

MATCHED EXAMPLE 17

$$X = \begin{bmatrix} \dfrac{7}{26} \\ -\dfrac{8}{13} \end{bmatrix}$$

CHAPTER 6

MATCHED EXAMPLE 1

9

MATCHED EXAMPLE 2

18

MATCHED EXAMPLE 3

28

MATCHED EXAMPLE 4

a The eight possible arrangements are RR, RB, RG, BR, BG, GR, GB, GG.

b There are five arrangements with a red marble.

MATCHED EXAMPLE 5

2520

MATCHED EXAMPLE 6

3 628 800

MATCHED EXAMPLE 7

a 336

b 840

MATCHED EXAMPLE 8

12

MATCHED EXAMPLE 9

10

MATCHED EXAMPLE 10

34 650

MATCHED EXAMPLE 11

48

MATCHED EXAMPLE 12

96

MATCHED EXAMPLE 13

126

MATCHED EXAMPLE 14

495

MATCHED EXAMPLE 15

133 784 560

MATCHED EXAMPLE 16

22 176

MATCHED EXAMPLE 17

$x = \{5\}$ $y = \{4,6\}$

MATCHED EXAMPLE 18

255

MATCHED EXAMPLE 19

3

MATCHED EXAMPLE 20

At least 35 books.

MATCHED EXAMPLE 21

At least 5 of the numbers will have the same remainder.

MATCHED EXAMPLE 22

Two numbers in the defined pigeonholes add to 12.

MATCHED EXAMPLE 23

The number of students who attend both guitar classes and piano classes is 12.

MATCHED EXAMPLE 24

There are 160 integers that are multiples of 2 or 7.

MATCHED EXAMPLE 25

There are 51 students who majored in mathematics, physics or computer science.

MATCHED EXAMPLE 26

There are 360 integers that are multiples of 3, 4 or 5.

CHAPTER 7

MATCHED EXAMPLE 1

a arc length = 9.42 units

b sector area = 56.55 units^2

c segment area = 5.64 units^2

MATCHED EXAMPLE 2

$a = 4.90$ cm

MATCHED EXAMPLE 3

$B \approx 65.39°$ or $114.61°$

MATCHED EXAMPLE 4

$x \approx 5.1$ cm

MATCHED EXAMPLE 5

$\theta \approx 44°$

MATCHED EXAMPLE 6

$Q \approx 76.5°$ or $103.5°$

MATCHED EXAMPLE 7

$YZ = 7.29$ cm

MATCHED EXAMPLE 8

$C = 57°$

MATCHED EXAMPLE 9

The dune is about 57.2 m high.

MATCHED EXAMPLE 10

The towers are about 7.21 km apart.

MATCHED EXAMPLE 11

The building is about 25 m high.

MATCHED EXAMPLE 12

The tower is about 1083 m high.

MATCHED EXAMPLE 13

a $\cot(60°) = \dfrac{1}{\sqrt{3}}$

b $\sec(30°) = \dfrac{2}{\sqrt{3}}$

c $\operatorname{cosec}(45°) = \sqrt{2}$

MATCHED EXAMPLE 14

$$\sec\left(\frac{4\pi}{3}\right) = -2$$

MATCHED EXAMPLE 15

If $\cot(x) = -\frac{5}{3}$, $\csc(x)$ for $\frac{\pi}{2} \le x \le \pi$ is $\frac{\sqrt{34}}{3}$.

MATCHED EXAMPLE 16

$$\tan(x + 3y) = \frac{\tan(x) + \tan(3y)}{1 - \tan(x)\tan(3y)}$$

MATCHED EXAMPLE 17

$$\sin\left(\frac{5\pi}{12}\right) = \frac{\sqrt{2} + \sqrt{6}}{4}$$

MATCHED EXAMPLE 18

$$\sin\left(\frac{\pi}{2} + x\right) = \cos(x)$$

MATCHED EXAMPLE 19

$$\cos(8x) = 2\cos^2(4x) - 1$$

MATCHED EXAMPLE 20

$$\sin\left(\frac{3\pi}{4}\right) = \frac{1}{\sqrt{2}}$$

MATCHED EXAMPLE 21

Proof: See worked solutions

MATCHED EXAMPLE 22

$$\cos(3x)\cos(5x) = \frac{1}{2}\left[\cos(2x) + \cos(8x)\right]$$

MATCHED EXAMPLE 23

$$2\sin(x) + 3\cos(x) = \sqrt{13}\sin\left(x + \tan^{-1}\left(\frac{3}{2}\right)\right)$$

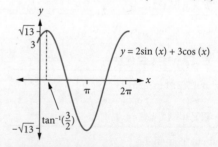

MATCHED EXAMPLE 24

$$\cos(x) - \sin(x) = \sqrt{2}\cos\left(x + \frac{\pi}{4}\right)$$

$$x = \frac{7\pi}{4} \text{ where } x \in [0, 2\pi]$$

MATCHED EXAMPLES 25–26

Proofs: See worked solutions

MATCHED EXAMPLE 27

$$a = \frac{\pi}{2}, \frac{3\pi}{2}$$

CHAPTER 8

MATCHED EXAMPLE 1

a $f: R \to R, f(x) = \log_2(x - 5)$ is a function.

The domain is $(5, \infty)$.

The range is R.

b It is not a function.

The domain is $[0, 6)$.

The range is $(2, 20]$.

c It is not a function.

The domain is $[-3, 7]$.

The range is $[-10, 0]$.

MATCHED EXAMPLE 2

a The basic function is $y = \sqrt[3]{x}$.

The function is dilated by the factor 5 parallel to the y-axis, reflected in the y-axis and translated 4 right and 1 up.

b The basic function is $y = x^4$.

The function is dilated by the by factor 0.5 parallel to the x-axis, reflected in the x-axis and translated 4 left and 2 down.

c $h(x) = \frac{2}{(x+2)} + 1$.

MATCHED EXAMPLE 3

a

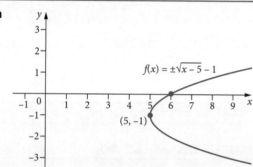

b

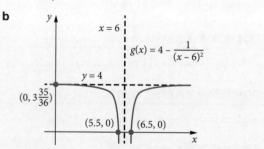

c

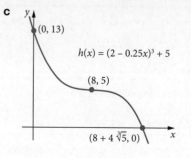

MATCHED EXAMPLE 4

$2x^2 + x - 2$

MATCHED EXAMPLE 5

$\dfrac{94}{x-4} + 5x + 24$

MATCHED EXAMPLE 6

$\dfrac{8x+7}{(x-1)(x+2)} = \dfrac{5}{x-1} + \dfrac{3}{x+2}$

MATCHED EXAMPLE 7

$\dfrac{2x^2+5x+1}{(x+2)^3} = \dfrac{2}{x+2} - \dfrac{1}{(x+2)^3} - \dfrac{3}{(x+2)^2}$

MATCHED EXAMPLE 8

$\dfrac{-10x^2-10x-8}{(x^2+4x+5)(x-1)} = \dfrac{-\dfrac{36}{5}x-6}{x^2+4x+5} - \dfrac{14}{5(x-1)}$

MATCHED EXAMPLE 9

a $\dfrac{8x^3+7x+4}{(x+5)(x+8)^2(x+7)} = \dfrac{A}{x+5} + \dfrac{B}{(x+8)^2} + \dfrac{C}{x+8} + \dfrac{D}{x+7}$

b $\dfrac{x^2+5}{(x+1)(5x^2+4x+2)(4x+7)} = \dfrac{A}{x+1} + \dfrac{Bx+C}{5x^2+4x+2} + \dfrac{D}{4x+7}$

c $\dfrac{x^2+2x+5}{(x+8)(x+4)} = A + \dfrac{B}{x+8} + \dfrac{C}{x+4}$

MATCHED EXAMPLE 10

$\dfrac{4x^2+4x-4}{(x-1)(x+1)^2} = \dfrac{3}{(x-1)} - \dfrac{2}{(x+1)^2} + \dfrac{1}{x+1}$

MATCHED EXAMPLE 11

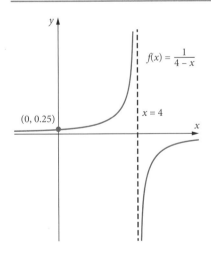

MATCHED EXAMPLE 12

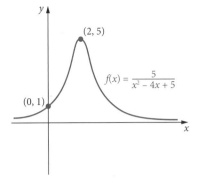

MATCHED EXAMPLE 13

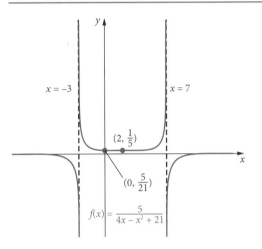

MATCHED EXAMPLE 14

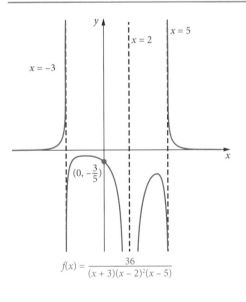

MATCHED EXAMPLE 15

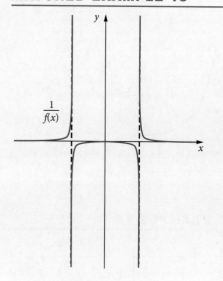

MATCHED EXAMPLE 16

a $\sec\left(\dfrac{\pi}{6}\right) = 1.154\ldots$

b $\operatorname{cosec}\left(\dfrac{5\pi}{4}\right) = -\sqrt{2}$

c $\cot\left(\dfrac{5\pi}{2}\right) = 0$

MATCHED EXAMPLE 17

$\sec(x) = -\sqrt{2}$ and $\cot(x) = 1$

MATCHED EXAMPLE 18

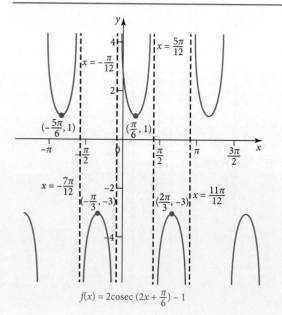

$f(x) = 2\operatorname{cosec}\left(2x + \dfrac{\pi}{6}\right) - 1$

MATCHED EXAMPLE 19

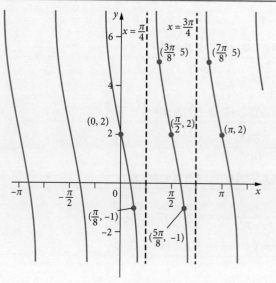

$y = 3\cot\left(2x - \dfrac{\pi}{2}\right) + 2$

MATCHED EXAMPLE 20

A possible equation is $y = 3\operatorname{cosec}\left(x + \dfrac{\pi}{6}\right) + 2$.

MATCHED EXAMPLE 21

a $\cos^{-1}\left(-\dfrac{\sqrt{3}}{2}\right) = \dfrac{5\pi}{6}$

b $\arctan\left(\dfrac{1}{\sqrt{3}}\right) = \dfrac{\pi}{6}$

c $\sin^{-1}\left(-\dfrac{1}{2}\right) = \dfrac{\pi}{6}$

MATCHED EXAMPLE 22

a The implied domain and range are $[-4, 2\pi + 3]$ and $[3, 0]$ respectively.

b The implied domain is $[0, 2]$ and the range is $(-\pi - 4, \pi - 4)$.

MATCHED EXAMPLE 23

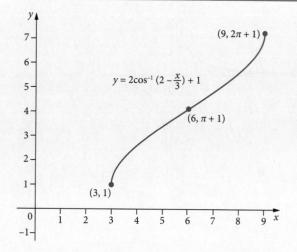

MATCHED EXAMPLE 24

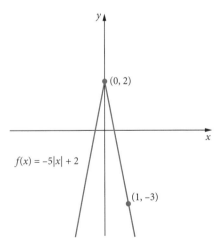

MATCHED EXAMPLE 25

a 25

b 54

c 16

MATCHED EXAMPLE 26

$m = -\dfrac{29}{10}$ or $m = -\dfrac{17}{2}$

MATCHED EXAMPLE 27

a $-\dfrac{5}{4} < x < 2$

b $x \geq 32$ or $x \leq -22$

c $x < \dfrac{1}{3}$ or $x > 1$

d There is no solution.

e $x < -\dfrac{3}{4}$

MATCHED EXAMPLE 28

a

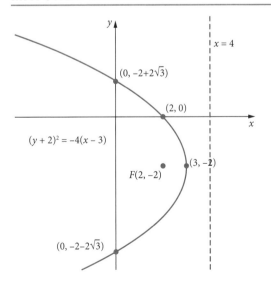

b

c

MATCHED EXAMPLE 29

The locus is the perpendicular bisector of AB, with equation $x - 4y - 13 = 0$.

MATCHED EXAMPLE 30

$(x - 1)^2 + (y - 3.5)^2 \approx 3^2$

The locus forms a circle with AB as a diameter, except the points A and B are excluded.

MATCHED EXAMPLE 31

The centre is $(2, -3)$ and the radius is $\sqrt{22}$.

MATCHED EXAMPLE 32

MATCHED EXAMPLE 33

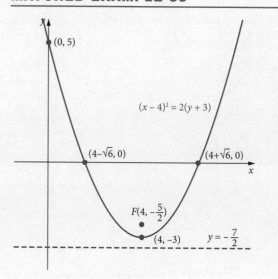

$(x-4)^2 = 2(y+3)$

$(0, 5)$

$(4-\sqrt{6}, 0)$ $(4+\sqrt{6}, 0)$

$F\left(4, -\frac{5}{2}\right)$

$(4, -3)$ $y = -\frac{7}{2}$

MATCHED EXAMPLE 34

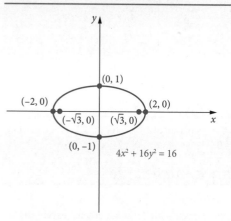

$(0, 1)$

$(-2, 0)$ $(2, 0)$

$(-\sqrt{3}, 0)$ $(\sqrt{3}, 0)$

$(0, -1)$ $4x^2 + 16y^2 = 16$

MATCHED EXAMPLE 35

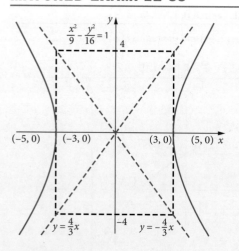

$\dfrac{x^2}{9} - \dfrac{y^2}{16} = 1$

4

$(-5, 0)$ $(-3, 0)$ $(3, 0)$ $(5, 0)$

$y = \frac{4}{3}x$ -4 $y = -\frac{4}{3}x$

MATCHED EXAMPLE 36

$\dfrac{(y+1)^2}{9} - \dfrac{4(x-3)^2}{9} = 1$

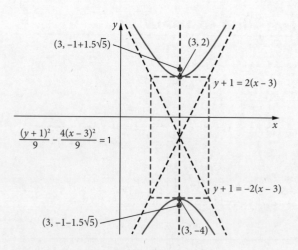

$(3, -1+1.5\sqrt{5})$ $(3, 2)$

$y + 1 = 2(x - 3)$

$\dfrac{(y+1)^2}{9} - \dfrac{4(x-3)^2}{9} = 1$

$y + 1 = -2(x - 3)$

$(3, -1-1.5\sqrt{5})$ $(3, -4)$

MATCHED EXAMPLE 37

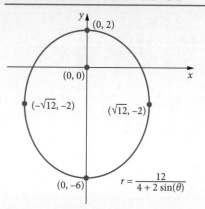

$(0, 2)$

$(0, 0)$

$(-\sqrt{12}, -2)$ $(\sqrt{12}, -2)$

$(0, -6)$ $r = \dfrac{12}{4 + 2\sin(\theta)}$

MATCHED EXAMPLE 38

The polar form is $r = \dfrac{16}{5 - 3\cos(\theta)}$.

MATCHED EXAMPLE 39

The parametric form is $\begin{cases} x = 6\cos(\theta) + 4 \\ y = 3\sin(\theta) - 1 \end{cases}$ for $0 \le \theta < 2\pi$.

MATCHED EXAMPLE 40

$(x - 2)^2 - \dfrac{y^2}{3} = 1$

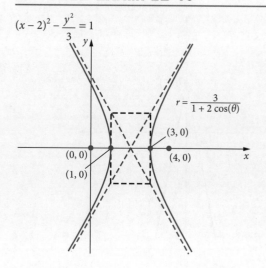

$r = \dfrac{3}{1 + 2\cos(\theta)}$

$(3, 0)$

$(0, 0)$ $(4, 0)$

$(1, 0)$

9780170464109

MATCHED EXAMPLE 41

The polar form is $r = \dfrac{7}{3 + 4\sin(\theta)}$.

MATCHED EXAMPLE 42

The Cartesian form is $\dfrac{x^2}{9} - \dfrac{y^2}{25} = 1$.

MATCHED EXAMPLE 43

$y^2 = 4(x + 1)$

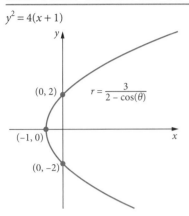

MATCHED EXAMPLE 44

$(y - 3)^2 = 8(x + 1)$

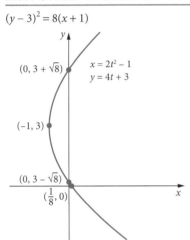

MATCHED EXAMPLE 45

$(x + 1)^2 + (y - 2)^2 = 16$

CHAPTER 9

MATCHED EXAMPLE 1

a $4i$

b $5i$

c $6\sqrt{2}i$

MATCHED EXAMPLE 2

a $-2 + 3i$

b $1 - 4i$

c $i^{\frac{1}{2}}$

MATCHED EXAMPLE 3

a $\mathrm{Re}(z) = 9$

  $\mathrm{Im}(z) = 6$

b $\mathrm{Re}(z) = -8$

  $\mathrm{Im}(z) = \sqrt{5}$

c $\mathrm{Re}(z) = 3x - 7y$

  $\mathrm{Im}(z) = 5x - 8y$

d $\mathrm{Re}(z) = \dfrac{7x}{x^3 - y^3}$

  $\mathrm{Im}(z) = \dfrac{4y}{x^3 - y^3}$

MATCHED EXAMPLE 4

a $4 - 5i$

b $10 - 20i$

c $-\dfrac{20i}{3}$

MATCHED EXAMPLE 5

a $-16 - 21i$

b $21 - 6i$

c $-175i$

MATCHED EXAMPLE 6

a $\bar{z} = -5 - 8i$

b $\bar{z} = -36 - 28i$

c $\bar{z} = \dfrac{1}{4} - 2i$

MATCHED EXAMPLE 7

a $-i$

b $-\dfrac{35}{65} + \dfrac{20i}{65}$

c $\dfrac{2}{3} - 2i$

MATCHED EXAMPLE 8

a $-29 - 34i$

b $9 - 42i$

c $-\dfrac{39}{41} - \dfrac{18}{41}i$

MATCHED EXAMPLE 9

$$\left(x-\frac{5}{4}-\frac{\sqrt{7}i}{4}\right)\left(x-\frac{5}{4}+\frac{\sqrt{7}i}{4}\right)$$

MATCHED EXAMPLE 10

a $\pm 5i$

The roots are purely imaginary.

b $2\pm\dfrac{\sqrt{6}}{2}$

The roots are real.

c $2\pm\dfrac{\sqrt{12}}{6}i$

The roots are complex.

MATCHED EXAMPLE 11

$b = -6$ and $c = 34$

MATCHED EXAMPLE 12

a

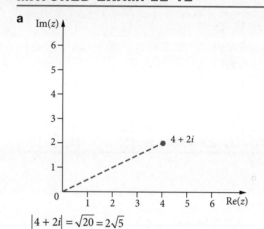

$|4 + 2i| = \sqrt{20} = 2\sqrt{5}$

b

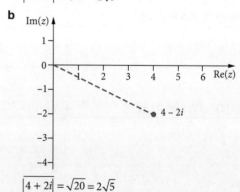

$|4 + 2i| = \sqrt{20} = 2\sqrt{5}$

c

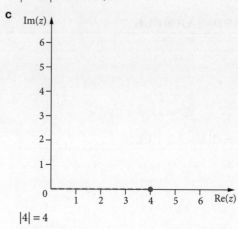

$|4| = 4$

d

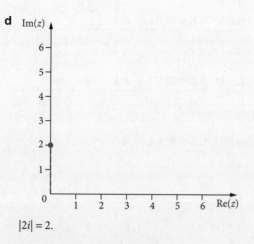

$|2i| = 2.$

MATCHED EXAMPLE 13

a 4

b $4\sqrt{2}$

c 5

MATCHED EXAMPLE 14

a $-3 + 7i$

b $3 - 7i$

MATCHED EXAMPLE 15

a $3\sqrt{2}\,\text{cis}\left(\dfrac{\pi}{4}\right)$

b $8\,\text{cis}\left(\dfrac{\pi}{3}\right)$

c $12\,\text{cis}\left(-\dfrac{5\pi}{6}\right)$

MATCHED EXAMPLE 16

a $-7i$

b $\dfrac{1}{\sqrt{6}}(-1+i)$

c $2 + 2\sqrt{3}i$

MATCHED EXAMPLE 17

a $53 - 9i$

b $\dfrac{3}{\sqrt{2}} + \dfrac{3}{\sqrt{2}}i$

c $-3.66 + 8.22i$

MATCHED EXAMPLE 18

a

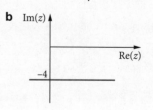

b

MATCHED EXAMPLE 19

a

b

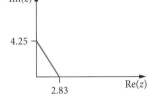

Angle = 124°

MATCHED EXAMPLE 20

a

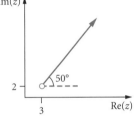

b

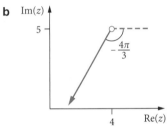

MATCHED EXAMPLE 21

a

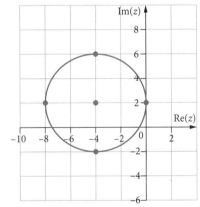

b $(x+4)^2+(y-2)^2=4^2$

The circle has radius 4 and its centre is at (−4, 2).

MATCHED EXAMPLE 22

a $|z-(-3+8i)|=9$

b $|z-(5+7i)|=8$

MATCHED EXAMPLE 23

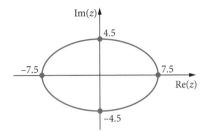

MATCHED EXAMPLE 24

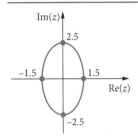

MATCHED EXAMPLE 25

a

b

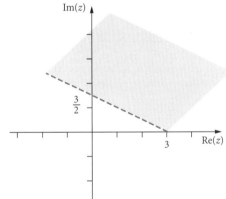

MATCHED EXAMPLE 26

a

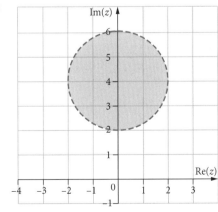

b

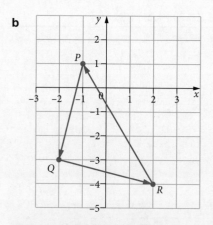

MATCHED EXAMPLE 27

a

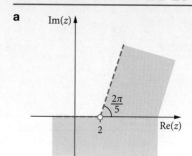

b

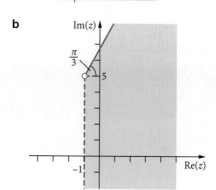

CHAPTER 10

MATCHED EXAMPLE 1

a

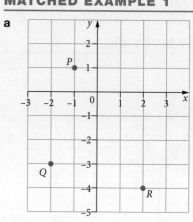

b

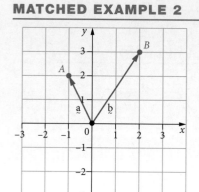

MATCHED EXAMPLE 2

MATCHED EXAMPLE 3

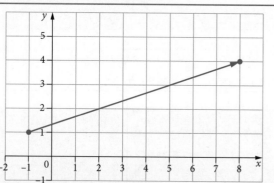

MATCHED EXAMPLE 4

a

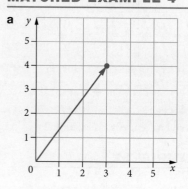

In the orthogonal unit form, this is $3i + 4j$.

b Magnitude = 5 units

Direction $= \tan^{-1}\left(\dfrac{4}{3}\right) \approx 53.13°$ from the positive direction of the x-axis.

MATCHED EXAMPLE 5

$$|\underset{\sim}{a}| = \sqrt{17}$$

$$\hat{\underset{\sim}{a}} = \frac{1}{\sqrt{17}}(\underset{\sim}{i} + 4\underset{\sim}{j})$$

MATCHED EXAMPLE 6

$\alpha \approx 53.13°$

$\beta \approx 36.86°$

MATCHED EXAMPLE 7

a $\underset{\sim}{u} + \underset{\sim}{v} = \begin{bmatrix} 1 \\ -6 \end{bmatrix}$ **b** $\underset{\sim}{u} - \underset{\sim}{v} = \begin{bmatrix} -5 \\ 4 \end{bmatrix}$

c $3\underset{\sim}{u} - 2\underset{\sim}{v} = \begin{bmatrix} -12 \\ 7 \end{bmatrix}$

MATCHED EXAMPLE 8

$$p = -\frac{13}{3} \text{ and } q = \frac{47}{9}$$

MATCHED EXAMPLE 9

$$\underset{\sim}{a} \cdot \underset{\sim}{b} = 7$$

MATCHED EXAMPLE 10

$$\underset{\sim}{a} \cdot \underset{\sim}{b} = 5\sqrt{3}$$

MATCHED EXAMPLE 11

$$\theta \approx 18$$

MATCHED EXAMPLE 12

a $\underset{\sim}{a} \cdot \underset{\sim}{b} = (3 \times 2) + (-1 \times 6) = 6 - 6$

$\therefore \underset{\sim}{a} \cdot \underset{\sim}{b} = 0$

b When $b = 9, a = 3\sqrt{3}$.

When $b = -9, a = -3\sqrt{3}$.

MATCHED EXAMPLE 13

$$\hat{\underset{\sim}{a}} = \frac{1}{\sqrt{53}}(2\underset{\sim}{i} + 7\underset{\sim}{j})$$

MATCHED EXAMPLE 14

Proof: See worked solutions

MATCHED EXAMPLE 15

The vector projection of $\underset{\sim}{a}$ in the direction of $\underset{\sim}{b}$ is $\frac{7}{2}(\underset{\sim}{i} + \underset{\sim}{j})$.

The perpendicular component of $\underset{\sim}{a}$ in the direction of $\underset{\sim}{b}$ is $\frac{-1}{2}\underset{\sim}{i} + \frac{1}{2}\underset{\sim}{j}$.

MATCHED EXAMPLE 16

The vector projection of $\underset{\sim}{a}$ parallel to $\underset{\sim}{b}$ is $4\underset{\sim}{i} - 2\underset{\sim}{j} + 2\underset{\sim}{k}$.

The vector projection of $\underset{\sim}{a}$ perpendicular to $\underset{\sim}{b}$ is $\underset{\sim}{i} + 3\underset{\sim}{j} + \underset{\sim}{k}$.

MATCHED EXAMPLE 17

The total force acting on the tractor is about 892 N at an angle of 52.3° to the horizontal.

MATCHED EXAMPLE 18

The change in velocity is about 17.7 knots at a bearing of 107.7°.

MATCHED EXAMPLE 19

Option E: $\underset{\sim}{e} = \underset{\sim}{a} + \underset{\sim}{d}$

MATCHED EXAMPLE 20

Proof

1 Sketch a diagram to set up the proof.

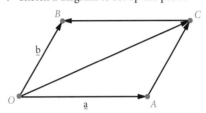

Let $\overrightarrow{OA} = \underset{\sim}{a}$ and $\overrightarrow{OB} = \underset{\sim}{b}$

2 State the proof step by step.

$\overrightarrow{OC} = \overrightarrow{OB} + \overrightarrow{BC} = \underset{\sim}{b} + \underset{\sim}{a}$

$\overrightarrow{BA} = \overrightarrow{OA} - \overrightarrow{OB} = \underset{\sim}{a} - \underset{\sim}{b}$

Consider

$\overrightarrow{BA} \cdot \overrightarrow{OC} = (\underset{\sim}{a} - \underset{\sim}{b}) \cdot (\underset{\sim}{b} + \underset{\sim}{a}) = \underset{\sim}{a} \cdot \underset{\sim}{b} + \underset{\sim}{a} \cdot \underset{\sim}{a} - \underset{\sim}{b} \cdot \underset{\sim}{b} - \underset{\sim}{b} \cdot \underset{\sim}{a}$

We know that the dot product is commutative and

$\underset{\sim}{a} \cdot \underset{\sim}{a} = |\underset{\sim}{a}|^2$

So, $\overrightarrow{BA} \cdot \overrightarrow{OC} = |\underset{\sim}{a}|^2 - |\underset{\sim}{b}|^2 = 0$

Since a rhombus has equal sides, $|\underset{\sim}{a}|^2 = |\underset{\sim}{b}|^2$

Thus, $\overrightarrow{BA} \cdot \overrightarrow{OC} = 0$.

3 State the conclusion.

The diagonals of a rhombus intersect at right angles.

CHAPTER 11

MATCHED EXAMPLE 1

x	3	4	5	6	7
$\Pr(X = x)$	$\frac{1}{6}$	$\frac{1}{6}$	$\frac{1}{3}$	$\frac{1}{6}$	$\frac{1}{6}$

MATCHED EXAMPLE 2

The mean of X is 1.

MATCHED EXAMPLE 3

a $E(X) = 1.4$

b $\text{Var}(X) = 1.64$

c $\text{SD}(X) = 1.280$

MATCHED EXAMPLE 4

a

n	20	40	60	80
$p(n)$	0.4	0.2	0.1	0.3

b **i** $E(N) = 46$

ii $\text{Var}(N) = 644$

MATCHED EXAMPLE 5

a $E(2Y) = 17$

b $Var(2Y) = 36$

c $SD(2Y) = 6$

MATCHED EXAMPLE 6

a The mean of X is 1.1 and the variance of X is 0.49.

b

T	0	1	2	3	4
$Pr(T = t)$	0.04	0.2	0.37	0.3	0.09

c The mean of T is 2.2 and the variance of T is 0.98.

MATCHED EXAMPLE 7

a $E(Y) = 7$

b $Var(Y) = 5$

c $SD(Y) = 2.24$

MATCHED EXAMPLE 8

The sample will not be random as the only people surveyed are the satisfied customers.

The shopkeeper has excluded an important subgroup of the population.

A better sample would be a random sample of the entire population for the city.

MATCHED EXAMPLE 9

7	109.7	30.0
11	113.3	26.6
19	96.0	23.6
20	120.3	38.9
27	118.3	35.71
28	109.1	30.2
32	100.9	27.3
33	114.1	32.8

MATCHED EXAMPLE 10

The proportion of sums greater than or equal to 6 is 0.5 or 50%.

MATCHED EXAMPLE 11

a The sample mean is 148.1.

b The value $\bar{x}$ is a sample statistic.

MATCHED EXAMPLE 12

a Event space = {(1, 1) (1, 2) (1, 3) (1, 4) (2, 1) (2, 2) (2, 3) (2, 4) (3, 1) (3, 2) (3, 3) (3, 4) (4, 1) (4, 2) (4, 3) (4, 4)}

b $A = \{(1, 4) (2, 4) (3, 4) (4, 1) (4, 2) (4, 3)\}$

MATCHED EXAMPLE 13

Brian wins two matches and John wins three matches.

MATCHED EXAMPLE 14

$\bar{x}$	$Pr(\bar{X} = \bar{x})$
1	$\dfrac{1}{16}$
1.5	$\dfrac{1}{8}$
2	$\dfrac{3}{16}$
2.5	$\dfrac{1}{4}$
3	$\dfrac{3}{16}$
3.5	$\dfrac{1}{8}$
4	$\dfrac{1}{16}$

MATCHED EXAMPLE 15

The probability of a mean greater than 70 is $\dfrac{3}{40}$.

MATCHED EXAMPLE 16

TI-Nspire

b   Mean of $\bar{x} = 119.975$

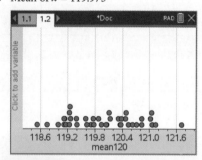

ClassPad

Mean of $\bar{x} = 120.4515$

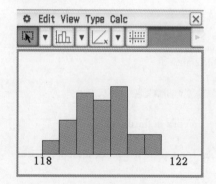

MATCHED EXAMPLE 17

Sample size	Mean of $\bar{x}$	Standard deviation of $\bar{x}$
40	79.9852	2.0936
110	80.006	1.1475

9780170464109

MATCHED EXAMPLE 18

The standard error is 0.490.

CHAPTER 12

MATCHED EXAMPLE 1

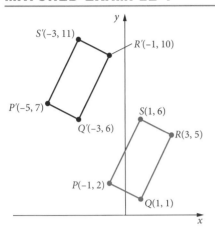

The image is $P'(-5, 7)$, $Q'(-3, 6)$, $R'(-1, 10)$, $S'(-3, 11)$. It is an identical rectangle.

MATCHED EXAMPLE 2

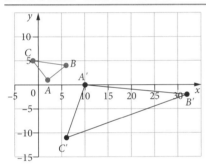

The image is $A'(10, 0)$, $B'(32, -2)$, $C'(6, -11)$. It is a scalene triangle.

MATCHED EXAMPLE 3

a The image is $K'(4, -1)$, $L'(0, 4)$, $M'(-1, 1)$.

b The image point is $(8, 6)$.

MATCHED EXAMPLE 4

The image is $A'(10, 0)$, $B'(32, -2)$, $C'(-6, -11)$.

MATCHED EXAMPLE 5

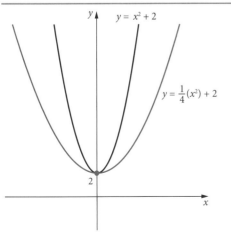

The image is $y = \dfrac{1}{4}x^2 + 2$; it is similar, not as steep, and its y-intercept is same as for $y = x^2 + 2$.

MATCHED EXAMPLE 6

The image is $A'(0, -2\sqrt{2})$, $B'(-2\sqrt{2}, 0)$, $C'(-2\sqrt{2}, -2\sqrt{2})$.

MATCHED EXAMPLE 7

The reflection is the straight line $y' = (2 - \sqrt{3})\, x' + 1 - \sqrt{3}$

MATCHED EXAMPLE 8

a The inverse is the dilation by the factor 0.8 from the x-axis.

b The dilation by the factor 1.25, dilated by the factor 0.8 gives the original point, so they are inverses

MATCHED EXAMPLE 9

The inverse reverses the transformation for $(4, -2)$.

MATCHED EXAMPLE 10

a $T \circ S(4, -3) = (-8, -18)$

b $S \circ T(4, -3) = (-2, -6)$

c $T \circ R(3, -4) = (-3, -4)$

d $R \circ T(4, -3) = (1, -5)$

MATCHED EXAMPLE 11

$T \circ S: (x, y) \to (4x + y, 11x + y)$

MATCHED EXAMPLE 12

$T \circ S: (x, y) \to (4x + y, 11x + y)$

MATCHED EXAMPLE 13

Proof: See worked solutions

MATCHED EXAMPLE 14

The gradients of the opposites were the same, so $ABCD$ is a parallelogram.

Also, AB is perpendicular to BC because the gradients are negative reciprocals of each other. Because these two segments are perpendicular, $\angle B$ must be a right angle. If $\angle B$ is a right angle, and the opposite angles are congruent, we get that $\angle C$ is also right and then it would follow that $\angle A$ and $\angle C$ must also be right angles. Because the parallelogram has four right angles. $ABCD$ must be a rectangle.

The area of the image of $ABCD$ under T is 30 square units.

MATCHED EXAMPLE 18

$R_2 \circ R_1 = \begin{bmatrix} \cos\theta & \sin\theta \\ \sin\theta & -\cos\theta \end{bmatrix}$, where $\theta = 180° + \tan^{-1}\left(\dfrac{3 + 4\sqrt{3}}{3\sqrt{3} - 4}\right)$.

The composition of reflection in the line $y = \tan(30°)$ and then in $y = 3x$ is equivalent to a rotation through the angle $180° + \tan^{-1}\left(\dfrac{3 + 4\sqrt{3}}{3\sqrt{3} - 4}\right)$.